THE SELFISH GENIUS

HOW RICHARD DAWKINS REWROTE DARWIN'S LEGACY

Fern Elsdon-Baker

Published in the UK in 2009 by
Icon Books Ltd, Omnibus Business Centre,
39–41 North Road, London N7 9DP
email: info@iconbooks.co.uk
www.iconbooks.co.uk

Sold in the UK, Europe, South Africa and Asia
by Faber & Faber Ltd, Bloomsbury House,
74–77 Great Russell Street,
London WC1B 3DA or their agents

Distributed in the UK, Europe, South Africa and Asia
by TBS Ltd, TBS Distribution Centre, Colchester Road
Frating Green, Colchester CO7 7DW

Published in Australia in 2009 by
Allen & Unwin Pty Ltd, PO Box 8500,
83 Alexander Street, Crows Nest, NSW 2065

Distributed in Canada by
Penguin Books Canada
90 Eglinton Avenue East, Suite 700,
Toronto, Ontario M4P 2YE

ISBN: 978-184831-049-0

Typeset in Minion by Marie Doherty

Printed and bound in the UK by
CPI Mackays, Chatham, ME5 8TD

CONTENTS

ACKNOWLEDGEMENTS

Thank you to all the countless people who have kindly offered advice, support and comment within my academic communities. I am especially indebted to Beth Hannon for her tremendous advice and contribution towards this book. I am also grateful to Matthew Conduct, Leucha Veneer and Frank James for comments on key sections. I would also like to thank Vicky Blake, Charlotte Nicklas, James Sumner and Mel Keene for their general humour, encouragement, advice and comments. I would also like to thank all my family and friends for their kind words, patience and the occasional beer.

Thank you also to Simon Flynn and Sarah Higgins for their fortitude, support, ideas and editing and to John Farndon for his helpful input and editing on earlier drafts.

I would like to thank all at the British Council, in particular those who work on the Darwin Now project, for their exceptional hard work and continued inspiration. I would also like to thank everyone I work with at both the British Science Association and the British Society for the History of Science for the opportunity and commitment to really put words into practice when it comes to communicating science and the history of science.

Last but by no means least I should like to thank James for everything.

ABOUT THE AUTHOR

Fern Elsdon-Baker is currently working with the British Council on the organisation of Darwin Now, an international project to celebrate the anniversary of Darwin's birth and the publication of his most famous book *On the Origin of Species*.

Having originally studied Environmental Sciences, Elsdon-Baker then went on to complete her doctoral studies in the history and philosophy of evolutionary theory. A passionate believer in the interactive communication of science and the history of science, in her spare time she also works in various voluntary roles within the British Society for the History of Science and the British Science Association.

INTRODUCTION

DARWIN'S ROTTWEILER

*Selfishness is not living as one wishes to live; it is asking others
to live as one wishes to live.*

Oscar Wilde

Let me begin this book by making a confession. The title of this
book, *The Selfish Genius*, is just a bit of mischief designed to
attract attention. I have no reason to think that Richard Dawkins,
the eponymous target of my joke, is any more selfish than he has
a perfect right to be. Nor for that matter – immensely clever and
influential as he is – does he necessarily fall, or should I say ascend,
into the realms of genius.

The title is of course a pun on the remarkable book that made
Dawkins' name, *The Selfish Gene*, and my choice of title was made
in, I might imagine, something of the same spirit as his. By 'selfish',
of course, as Dawkins made clear, he was not intending to imbue
genes with dubious moral qualities. He was simply using a neat
and eye-catching shorthand to describe the way in which genes
might inevitably interact with the world – with no intention, selfish or otherwise.* It is the nature of genes, Dawkins argued, to act
entirely for their own survival, and this is what drives the evolution of life. It's not that they choose to; they just do. Every living
organism, Dawkins contended, is simply a vehicle built entirely by
and for the survival of its pertinacious genes.

---

* Actually, I believe that Dawkins intended it as a metaphor as well – and
wanted to suggest that it was quite helpful to think of a gene as if it did have a
mind with selfish motives.

1

It is in the same spirit that I use the word 'selfish' in the title. It is my contention that Dawkins' approach to evolutionary science, intentionally or otherwise, has come to dominate the field unhealthily – or rather to dominate the public perception of evolutionary science. In this year of Darwin anniversaries, 200 years after Darwin's birth and 150 after the publication of *On the Origin of Species*, it is Dawkins who is Professor Evolution, the pundit called in to give the definitive word on Darwin for the layman. And, of course, when it comes to the attacks of creationists on evolution, it is Dawkins who is out there fronting the battle against them, waving the torch of rational science, so much so that in many people's eyes he is not just Professor Evolution, but Professor Science too.

Letting out the genie

What I'd like to do in this book is not to diminish Dawkins' impressive contribution to the communication of evolutionary science, but rather to put it in context and redress the balance that has been upset, partly by the sheer force of Dawkins' personality and his indisputably brilliant way with words. It is my belief that Dawkins has effectively hijacked Darwin and distorted his legacy to champion an inflexible approach that gives the public a very one-sided view of what's really going on in evolutionary science.

Moreover, he has gradually opened up his attack on creationists, who oppose the very idea of evolution, to target the entire realm of religious belief in his best-selling book *The God Delusion*. As he has done so, he has presented himself as standing at the vanguard of clear-thinking rational science, fighting against a swelling tide of fanatical delusionists. The effect has been to draw science and religion into a knockabout fight that does service to neither, and certainly reduces the chances of calm, rational debate.

I should state, right from the outset, that I am actually on Dawkins' side, or rather on the side of science. I am no undercover creationist trying to sneak an attack on true science under the guise of pseudoscience. I come from a secular background, and I was brought up in a household of scientists. Even our childhood pets were named after scientists!

The point here is that scientific debate and discussion have always been a part of my life, and they always will be. So I'm certainly not weighing in on the side of those who would give evolution short shrift. No, I am firmly in the scientific camp, and I wholeheartedly think that Darwin's elucidation of his theories of evolution was one of the most important scientific breakthroughs in the history of science.

The origins of neo-Darwinism

Dawkins is often described as a neo-Darwinian, and I think I should briefly explain just what is meant by that, though I shall be coming back to it later in more detail. The orthodox, or rather the popular, version of the progress of evolutionary science goes as follows.

In 1859, after more than two decades of diligent research, Charles Darwin published his breakthrough book *On the Origin of Species*. In it he expounded the theory, announced simultaneously the previous year by Darwin and Alfred Russel Wallace, that all the wonderful, teeming variety of life on earth has descended from common ancestors, and that all the multitude of different species have evolved gradually like an ever-branching tree, as natural variations have been accentuated over time by a process he called 'natural selection'.

Natural selection was a devastatingly simple idea. No two organisms ever come into the world quite alike. Occasionally, a slight difference gives an organism a crucial edge in the prevailing conditions – a better chance of surviving longer and producing more progeny. And as it and its descendants successfully produce more progeny, others which are less well adapted to the prevailing conditions die out – a process vividly dubbed by Darwin's contemporary Herbert Spencer as 'survival of the fittest'.*

Darwin stalled

There were two key elements to Darwin's theory – first that life had evolved gradually, and second that it had evolved by means of natural selection. Despite the famous furore provoked by Darwin's books, the rumpus over evolution actually died down quite quickly and evolution as an idea was soon quite widely accepted, not only among naturalists but by all but the most diehard clergymen, too. 'Evolution' was by no means a new idea, as we'll see later, and the sheer weight of evidence that Darwin presented, in his careful, understated way, was just too hard to deny. Ironically, the opposition of creationists today is, in some ways, more organised and sustained now than it was 150 years ago – despite the overwhelming weight of evidence for it that has emerged since.

However, it was not evolution but Darwin's mechanism for evolution, natural selection, that took a long time to gain widespread acceptance. One problem that is often cited to explain this is that Darwin just hadn't provided evidence showing how the inheritance of beneficial traits actually happened, or how these

---

* Although 'survival of the fittest' is now popular shorthand for Darwin's theory, many biologists avoid it. That's because now for some biologists 'the fittest' technically means the best able to survive, so the phrase is a tautology – 'the survivors survive'. In *The Extended Phenotype*, Dawkins explains in some detail how he thinks this apparent tautology arose, and how some academics have mistakenly used it to argue that the whole idea of natural selection is a logical tautology rather than just this particular phrase.

differences might be sustained and strengthened through the generations – rather than just getting gradually lost as they blended in with the general population each time they emerged. But, as we shall see, this version of events is not quite the whole story.

Soon, natural selection was left in the hands of just a few 'neo-Darwinists' such as Wallace to champion. If only poor Darwin had known, the popular story goes, that the answer to his dilemma was actually out there waiting for him all the time, just across Europe in the work of the Austrian monk Johann 'Gregor' Mendel. It was Mendel whose groundbreaking experiments with pea plants in the monastery garden in the 1850s laid the foundations of genetics, as he showed how traits can both be preserved intact from generation to generation and can skip generations, through the combination factors, later called genes, inherited from each parent. Though Mendel was far from being isolated from the science community, his work attracted little attention at the time and Darwin never got to know about it. As a result, the neo-Darwinians were fighting an uphill battle to sustain the theory of natural selection, even though evolution quickly became orthodox.

Gene selection

In the early twentieth century, Mendel's work was rediscovered and eventually made a contribution to the development of modern classical genetics. For a while, the either/or, stop-start nature of genetic inheritance seemed at odds with the gradualism that was needed to explain evolution by natural selection. Dawkins has evocatively described genetic inheritance as a 'digital' process that at first seemed to conflict with the 'analogue' idea then imagined for evolution. Then, in the 1920s and 30s, work by J.B.S. Haldane, Sewall Wright and Ronald Aylmer Fisher sought a resolution.

With this reconciliation, biologists could at last dovetail genetics and evolution by natural selection together into what came to be called by some the 'modern synthesis' and by others, including Dawkins, the 'neo-Darwinian synthesis'. Dawkins was clearly heavily influenced by these thinkers, declaring in the 1989 edition of *The Selfish Gene* that 'the gene's eye view of Darwinism is implicit in the writings of R.A. Fisher and the other great pioneers of neo-Darwinism in the early thirties'.

The synthesis seemed so logical and powerful that soon natural selection of genetic variations was seen as the only possible method of evolution, a shift later described by a famous American palaeontologist, the late Stephen Jay Gould, as a 'hardening of the modern synthesis'. It was hardened further still in the 1960s during arguments about the level on which selection actually took place – that is, whether it was groups within species, individual organisms or just genes that were actually being selected.

For goodness' sake

Altruism had always seemed to be a problem for the supercompetitive survival image of natural selection evolution. When a bird cries out to warn its fellows of danger, for instance, it exposes itself to extra risk. How could such behaviour – which actually reduces the individual bird's survival chances – ever have evolved by natural selection, which roots out those less likely to survive? 'Group' selectionists had long assumed that it was all about what was for the 'good of the species' – i.e. these were traits which helped the group or the species survive, if not the individual. Then, in the mid-1960s, George Williams brutally exposed what some saw as huge flaws in this logic.

The floodgates opened, and very quickly evolutionary biologists such as the late Bill Hamilton were saying that it is neither species nor organisms that are selected, but individual genes. They talked, for instance, about altruism being about 'kin selection' – which is how genes that programme an organism to act for the good of its relatives could evolve. This is where Dawkins came in.

Dawkins arrives

In 1976, Dawkins dramatically unleashed these arguments on the general public and took them to their logical extreme in his bestselling *The Selfish Gene*. There is no doubt that Dawkins' book was a landmark. It was partly a synthesis of then-current ideas on gene selection, but it was so clearly and accessibly written, and so cogently argued, that it became much more than that. Some people have described it as the most influential science book of the twentieth century, and many biologists have commented on what an extraordinary effect it had on them.

In a special event held in 2006 to commemorate the 30th anniversary of its publication, the philosopher Daniel Dennett said: 'When I read the book, it changed my life.' And Dennett was not alone. Its impact on the public, especially the impact of its title, was equally profound. It gave evolutionary biology a new and exciting focus, in searching for the links between genes and adaptations, and it gave the public a powerful and mechanistic new image of evolution at work.

Interestingly, the book's timing was apposite, coming at a time when the feelgood 1960s were giving way to the rampant individualism of Mrs Thatcher's 1980s, exemplified by her infamous comment, 'There is no such thing as society'. For many people Dawkins' title, *The Selfish Gene*, seemed to symbolise or even justify that shift, implying that, even right down to our genes, we are selfish and it makes sense to recognise this. Some even used it as a justification for nakedly self-interested behaviour, saying 'It's all in our genes, anyway'. Dawkins has since expressed his regret that his title could be interpreted in this way, but the 'selfish gene' genie was by then out of the bottle.

Dawkins positioned himself, then as he does now, as a neo-Darwinian. He viewed *The Selfish Gene*'s central argument – that evolution is all about the natural selection of genes, and genes alone – as taking Darwin's theory of evolution by natural selection to its logical conclusion. *The Selfish Gene* is presented as purified and concentrated Darwinism, reduced to its very essence. That uncompromising logic has always been a strong element in Dawkins' work. But all science is about empirical observation as well as logic, and what appears to be the logical conclusion is not necessarily the right one.

Dawkins versus Gould

Such has been the impact of *The Selfish Gene* that it is not so widely known outside academic circles that Dawkins has always had his scientific, as well as religious, opponents. The most famous was the American palaeontologist Stephen Jay Gould, who died in 2002. For decades, a battle royal was waged between Dawkins and Gould, with intellectual heavyweight supporters joining in on both sides. So prolonged and so polemical were the exchanges – and they are still going on between Dawkins' and Gould's respective supporters – that it has created a major chasm in the intellectual community.

On the surface, it seems remarkable that two such staunch defenders of science and evolutionary theory against creationists and religious zealots should be so irreconcilably opposed – both featured in this role in *The Simpsons* (Gould) and *South Park*

(Dawkins).* Dawkins argues that gradual adaptations created by natural selection of genes are at the heart of evolution, and that natural selection drives evolution. Gould, on the other hand, believed from his study of fossils that evolution is much more intermittent and centres on changes in species rather than genes, with natural selection playing only a part, rather than controlling the entire process. Much of what Gould called 'adaptationism' – explanations of adaptations by Dawkins and his supporters – he regarded as no more credible than Kipling's *Just So Stories*.

These are indeed crucial scientific differences, but at the heart of the argument, too, lay fundamentally opposed ideas about science and how it should be practised, on which I shall expand during the course of this book. Dawkins believed that Gould's thinking was woolly, while Gould labelled Dawkins a 'Darwinian fundamentalist' – as dogmatic, he believed, as the religious fundamentalists Dawkins frequently lambasts.

The molecules of life

Gould and his supporters, though, were very much on the back foot in the 1980s and 1990s, as scientists with powerful microscopes began to probe into what seemed to be the very secrets of life at the molecular level. Where was the relevance of Gould's macro-view of life when down at Dawkins' micro-level genes were capturing all the headlines? Not only were molecular biologists unravelling the genetic code in astonishing detail, but genetic engineers were also learning how to manipulate the code to alter

---

* It is perhaps telling that Gould should feature in the subtler satire of *The Simpsons* in his mild-mannered and professorial way, while Dawkins should find himself quite at home in the mouthy, brash *South Park*, taking over as South Park Elementary's evolution teacher in his characteristic shoot-from-the-hip manner.

life directly. Breakthrough followed breakthrough with the first GM (genetically modified) animals, then GM crops, and then the cloning of mammals. Each year expectations were being raised that scientists would one day be able to eliminate all hereditary diseases using gene therapy. Investment in biotech companies went through the roof and funding in molecular biology research mushroomed.

The culminating triumph of this molecular revolution was the mapping of the entire human genome – the complete sequence of genes on human DNA – in the early years of this century. The first map of the genome took years to complete, as even a short sequence of bases took months to decipher. Now there are machines that can pick up fluorescent tags to read a million of the chemical bases that make up the DNA code in just a second. Many kinds of plants and animals besides humans have already had their genomes mapped, and soon machines will be able to map the entire genome of any person who wants it in a matter of days (at just $5,000 a time). In his recent book, *Darwin's Island*, professor of genetics Steve Jones described the meaning for evolution of this scientific revolution thus: 'A science that had been in its infancy, a mere description of bones and muscles, became an adolescent when *The Origin of Species* showed how shared structure was evidence of common descent. It has at last matured. The anatomy of DNA has become the key to the history of life.'

Losing our genes

Yet this explosion of molecular research in recent years has thrown up some crucial surprises. First of all, the role of genes has turned out to be far more ambiguous than previously imagined. Scientists had thought, for example, that the size of the genome would bear a direct relationship to the complexity of the organism. So, fantastically complex humans would have fantastically big genomes. But the completion of the human genome has really cut human gene aspirations down to size. It turns out that instead of the hundreds of thousands of genes estimated for humans, we have around a mere 20,000–25,000 – about the same as a mouse! In the post-genomic era our concept of how evolution might work has begun to change fundamentally. Scientists are having to dramatically rethink just what it is that genes do, and what part they play in evolution.

Dawkins, of course, has never said that genes were all that drove evolution. He talks instead about the as yet abstract idea of 'replicators' – basic evolutionary units – and describes genes, or rather sets of genes, as being the most likely candidates to be those replicators. In *The Extended Phenotype*, he also proposed that so-called 'junk' DNA was a product of DNA trying to replicate *itself* just as much as the organism. Yet the most recent empirical research is blurring and complicating the apparently inexorable logic of the selfish gene vision, as we shall see later.

The shift has been so dramatic that I do not know of many practising evolutionary biologists who are now committed 'gene-centrists' in the Dawkins mould – and the unambiguous picture often presented to the public does less than justice to the exciting ways in which working scientists are continuously making discoveries that don't always confirm old theories, but sometimes throw them into doubt. Wonderful, coherent theories make great books and great TV but don't always correspond to the messy, confusing business of science in practice.

Lamarck's brothers (and sisters)
In fact, studying nature under the microscope has thrown up what might just turn out to be even more fundamental challenges

to neo-Darwinian purists like Dawkins. Half a century before Darwin published his ideas, the French naturalist Jean-Baptiste Lamarck had come up with a forerunner to Darwin's theory of evolution which depended not upon natural selection of variations with which organisms start life, but on the inheritance of traits acquired during their lifetime in response to their surroundings – the hoary classic example of which was the giraffe, which was said to have acquired its long neck by stretching for higher branches. Organisms evolved, in the Lamarckian system, by passing on such traits to their offspring, a process later known as the inheritance of acquired characteristics, or Lamarckian inheritance.

Neo-Darwinists have always ridiculed this idea ever since the leading light of the first generation of neo-Darwinists, August Weismann, did an experiment in which he cut off the tails of mice to show that their offspring weren't born tailless too, showing – so the story goes – that they clearly didn't inherit acquired characteristics! But, as we shall see later, this inference is perhaps not as clear-cut as it first seems. In pre-DNA times, 'Weismann's barrier' was said to prevent changes acquired during an organism's life from ever crossing into the sex cells, the cells from which offspring develop. So evolution could only take place through mutations in the heritable material of sex cells – what we would later think of as genetic material.

Following the discovery of DNA, neo-Darwinists talked about the 'central dogma' of molecular biology introduced by Francis Crick in 1958 which, in simple terms, said that though information could be transferred from DNA to protein, it could not be transferred from protein to DNA – which effectively rules out the possibility of DNA being altered by life's events. Dawkins has said, in his book *The Extended Phenotype*, 'to be painfully honest,

I can think of few things that would more devastate my world view than a demonstrated need to return to the theory of evolution that is traditionally attributed to Lamarck. It is one of the few contingencies for which I might offer to eat my hat.'

There has always been a contingent of evolutionary scientists who believed that the dogmatic neo-Darwinian line was misplaced, but the progress of genetic research seemed to put them on the defensive. In recent years, though, research has actually given so-called 'neo-Lamarckists' succour. While it is not yet time for Dawkins to start softening up his headgear, research by a number of scientists has shown that there may be other ways of passing on characteristics apart from via the genes, and that these other ways suggest it is quite plausible that changes that happen during an organism's lifetime could be passed on.

Knocked sideways

Moreover, bacteria and other micro-organisms were until recently largely left out of the evolutionary picture, but in recent years improvements in microscopy and gene analysis have made them increasingly popular subjects for study. This is because they multiply so quickly that scientists can see adaptations that might take tens of thousands of years to evolve in mammals happening right before their eyes. Studies of bacteria have shown that 'horizontal gene transfer' – the direct transfer of genes from one organism to another – is far more common than anyone previously imagined. Further studies have shown that such transfers might happen even among higher animals. And if organisms can swap genes so readily, the strict unidirectional inheritance of characteristics demanded by the logic of the neo-Darwinians is thrown into doubt.

One of the problems with all these challenges to the neo-Darwinian orthodoxy, as Dawkins has been quick to point out, is that creationists leap on them as tools to attack the whole idea of evolution. There you are, the creationists say triumphantly: Darwin was wrong, and the cracks are at last appearing! Of course, they are not cracks in the theory of evolution at all; they are only cracks in neo-Darwinian purist theory. In fact, they are simply deepening and refining our ideas of how evolution works. What is more, as I will show in this book, the neo-Darwinians have actually presented Darwin's ideas in a way that best suits their perspective, and his thinking encompassed these new developments far better than they would have us believe.

I can imagine that many – perhaps including Dawkins himself – would say that scientists need to present a united front against the religious extremists, especially in this age when fundamentalism, it is argued, seems to be gaining footholds in both the Christian and Islamic worlds. I would counter that scientists should never dance to the fundamentalists' tune. They should never present themselves as certain when there is doubt. The very success and truthfulness of science is founded on doubt and scepticism. It moves forward by continually rethinking, reobserving and rechecking against reality again and again to expose the flaws in current ideas. That is, I believe, at the heart of science's claim to knowledge – because it continually tests its ideas to the limits, and never just doggedly defends them.

Dawkish behaviour

Over the years, like a successful coloniser, Dawkins has gradually extended his territory. His research background is in ethology, and so it is not surprising that in *The Selfish Gene* he extended his theory of the selfish gene to cover not just physical traits, but also the evolution of behaviour. For him, it was only logical that if animals had evolved, so must their behaviour, and gene selection has been a very successful way of explaining some otherwise inexplicable behaviour patterns in animals, especially the social insects such as bees and ants.

Of course if the selfish gene is relevant to animal behaviour, it must be relevant to human behaviour too. In fact, much of Dawkins' book was devoted to human behaviour, and this is where he was at his most controversial and also his most speculative. He brought little evidence, beyond the anecdotal, to bear in his development of ideas which were given snappy titles like 'Genesmanship', 'You scratch my back, and I'll ride on yours' and 'Nice guys finish first'. The logic was persuasive and the argument fascinating, but it was basically conjectural.

In his countless forays beyond the boundaries of his native ethology, Dawkins is sometimes right and sometimes wrong. His arguments are occasionally speculative and some are more or less unprovable one way or the other, at least at present. But one criticism I would make is that his speculation has no obvious boundaries, so it is always unclear what is dispassionate, clear-sighted scientific speculation and what simply represents Dawkins' own personal and political agenda. And I would argue that this does not in any way constitute objective science.

I imagine Dawkins might counter that he is merely setting the ball in motion, throwing out ideas for scientists and other academics to follow up in a proper scientific way. Certainly his biggest speculative idea about human behaviour and culture, the idea of 'memes', has inspired a deluge of research and opinion. Although humans have risen above the biological power of the natural selection of their genes to direct their behaviour, Dawkins argued, human culture is subject to the same process of evolution by natural, or rather cultural, selection – only the unit being selected is not a gene or set of genes but what he dubbed a 'meme'. A meme could be a song, a catchphrase or a religious belief, for example, and like a gene it spreads and survives or dies out according to how it mutates.

I am not too concerned about the quality and validity of Dawkins' ideas. That is for those scientists who are interested to support or disprove with real evidence. What does worry me is the style of presentation. In both his books and his TV shows, Dawkins uses a highly charged, political style of passionate advocacy to promote his arguments. This is certainly eye-catching and persuasive, and helps sell books by the truckload, but it is hardly the approach of a dispassionate scientist.

I am all for communicating science to the public both clearly and effectively. Indeed, that is my current job, as I take time out from academic research to work on the Darwin anniversary celebrations. I am also all for scientists caring deeply about their work. Dawkins is undoubtedly a brilliant communicator, and cares deeply about what he says. But scientists also need to be as measured and objective in what they present to the world as they try to be in their research, otherwise what they say can be highly misleading.

Dawkins' official position between 1995 and 2008 was Charles Simonyi Professor for the Public Understanding of Science at Oxford University, but for me his narrow advocacy of just one of the many avenues of thinking on evolution, and his often speculative, combative approach, may actually obscure public understanding of the real issues, not enhance them. My worry, too, is that such is his extraordinary prominence that his singular vision may actually act to constrain debate in other fields of evolutionary biology. Moreover, as I said earlier, I believe that such an adversarial style only adds fuel to the feud between the two warring parties of 'Darwinism' and 'creationism' in a way which is detrimental to science.

Stretching the envelope

Interestingly, it is in the opening of his most academic book, *The Extended Phenotype*, that Dawkins acknowledges his approach most plainly, stating: 'This is a work of unabashed advocacy ... What I am advocating is not a new theory, not a hypothesis which can be verified or falsified, not a model which can be judged by its predictions.' For someone who argues so vociferously on behalf of science, it seems strange that he should be willing to follow a line like this that runs contrary to every accepted scientific value.

Yet ironically, despite this admission, or should I say challenge, *The Extended Phenotype* is actually Dawkins' most successful and measured work – maybe the thought of his stated scientific audience encouraged a restraint that has not always been present in his more populist works. In this book Dawkins took the selfish gene argument a step further, and tried to answer some of the criticisms of his earlier work.*

* In particular, Dawkins countered the accusation that he is a genetic determinist – that his selfish gene implies that genes inevitably determine the development of every physical feature and every form of behaviour. Psychologist Steven Rose joked that genetic determinism gave philandering males a great excuse: 'Don't blame your mates for sleeping around, ladies, it's not their fault they are genetically determined.' But Dawkins argues that although genes are selected and survive through their effects, this does not mean that they determine an organism's entire life. A weakly built man can build up his body through exercise, and while there may actually be a 'gene for reading', it doesn't mean that we don't actually have to put in a huge effort to learn to read. 'A sensible and unexceptionable way of thinking about natural selection – "genetic selectionism"', Dawkins says, 'is mistaken for a strong belief about development – "genetic determinism".'

The phenotype is defined as the visible characteristics of an individual organism – its anatomy, biochemical processes, behaviour and so on. Dawkins argued in *The Extended Phenotype* that in their drive for survival and replication, genes do not merely shape the phenotype of an individual organism, but extend their influence into the world around it when it also affects their chances of survival. He cites a beaver dam as an example of such an extension of the phenotype. According to Dawkins, the 'central theorem' of *The Extended Phenotype* is that 'An animal's behaviour tends to maximise the survival of the genes "for" that behaviour, whether or not those genes happen to be in the body of the particular animal performing it'.

However, crucially, Dawkins asserted that genes or genotype (the organism's assembly of genes) can have an impact on the environment via the organism or the phenotype – but that the environment can never affect the genotype via the phenotype. If it were ever shown that the environment could have an impact on the genotype, then Dawkins' singular vision of the selfish gene might be seriously weakened. And it is this possibility which I would argue is Dawkins' real *bête noire*, rather than religion. Theology could never disprove his 'selfish gene hypothesis', but science just might.

Climbing Mount Dawkins

After *The Extended Phenotype*, Dawkins was less and less concerned with developing evolutionary science, and more interested in the logical and philosophical implications of his ideas. In *The Blind Watchmaker*, for instance, he argued fiercely against the notion that awesome complexity and apparent ingenuity in nature must imply that it was deliberately designed rather than being simply the mechanical outcome of evolution. In *Climbing*

Mount Improbable, he elaborated on this to show how organisms can evolve to a peak of astounding and improbable complexity as chance mutations are given direction simply by the blind operation of natural selection – not by any guiding hand or any notion of where they might be heading.

My own first encounter with Dawkins was as an undergraduate, reading *The Blind Watchmaker*. Dawkins' title comes from an anecdote in *Natural Theology*, a book dating from 1802 by an Anglican clergyman called William Paley, in which Paley wrote:

In crossing a heath, suppose I pitched my foot against a stone, and were asked how the stone came to be there; I might possibly answer, that, for anything I knew to the contrary, it had lain there for ever: nor would it perhaps be very easy to show the absurdity of this answer. But suppose I had found a watch upon the ground, and it should be inquired how the watch happened to be in that place; I should hardly think of the answer which I had before given, that for anything I knew, the watch might have always been there.

Dawkins suggested that Paley 'gave the traditional religious answer to the riddle [that the complexity of the watch implied a watchmaker – and so the complexity of nature must imply a divine watchmaker.] ... The true explanation is utterly different, and it had to wait for one of the most revolutionary thinkers of all time, Charles Darwin.'

This narrative of the brilliant atheist Darwin driving away the darkness of an earlier age of creationist ignorance sounds so utterly compelling and believable, and fits so neatly into the popular story of the rise of Darwinism that I outlined earlier, that most readers have taken it as a true historical account. But in my eager undergraduate way back then, I did a little research. Then I started to see that Dawkins' version of history wasn't altogether accurate – I found it so oversimplified, indeed, that I made it the subject of my PhD in order to explore what really happened! That was when I began to discover that the whole neo-Darwinist version of the history of evolutionary science is actually rather a one-sided affair, as I shall show in the first section of this book.

Science Wars

Does this matter? Surely it's all in the past, and any quibbling would simply be the nit-picking of an academic mouse, nibbling at the feet of the intellectual giants? Well, actually, I think it does. Of course, I could grandly say that truth always matters, but that would be pretty vacuous. No, what does matter is that this 'mythical' version of the past could be diverting attention from the real issues in evolutionary science, and perpetuating the existence of 'opposing camps' where there is no need for them. Indeed, I feel it may be helping to sustain an orthodoxy that, if not actually hindering the progress of evolutionary science (which it just might – orthodoxies often do), helps to skew understanding of Darwinism outside academic circles.

I think, too, that this is why I want to question just which idea of science Dawkins truly wishes to champion. Little noticed by the outside world, a ferocious war has been waged in the groves of academe about just what science is. The issues are deep and complex, and I'll explore them in more detail later, but they began more than 100 years ago as 'scientists' sought to professionalise their areas of study. It developed into a clash between academics in the humanities and scientists, characterised by C.P. Snow in the middle of the last century as the 'Two Cultures' so evident in the arts/science split, and gained extra momentum when Thomas Kuhn suggested in the 1960s that science is not entirely empirical and that theories might change with shifts in scientific culture. Out of that grew a new academic discipline called Science and Technology Studies which sought to study the sociology of science. The dispute reached a climax with the so-called 'science wars' of the 1990s, in which purist scientists fought tenaciously for science's claim to the truth against those whom they scornfully called 'relativists'.

Dawkins was and is very much on the purist side, snorting derisively in *River Out of Eden*: 'Show me a cultural relativist at 30,000 feet and I'll show you a hypocrite ... If you're flying to an international conference of anthropologists or literary critics the reason you will probably get there ... is that a lot of Western scientifically trained engineers have got their sums right.' This kind of shoot-first-ask-questions-later put-down certainly pleases some punters and makes a great quote, but it trivialises a serious debate about the real nature of science and the kinds of answers it provides. It also presumes that this is a dichotomous debate with no room for middle ground between the 'sciences' and the 'humanities'.

My feeling is that this too helps to distort the public understanding about the nature of science in the wider world – and actually hands valuable ammunition to the creationists and the religious fundamentalists who are the target of Dawkins' scorn in *The God Delusion*, the best-seller in which he notoriously derides all religion as being fundamentally false. I am a great admirer of Dawkins, and I believe that his contribution to evolutionary science has been a very valuable one, but I think it is vitally important to put his work in context, or we do science a great disservice – and, ironically, disguise the real value of Dawkins' own achievement. I will do what I can to redress the balance in the remainder of this book.

PART I

CHAPTER 1

THE ORIGINS OF EVOLUTION

Organic life beneath the shoreless waves
Was born and nurs'd in ocean's pearly caves;
First forms minute, unseen by spheric glass,
Move on the mud, or pierce the watery mass;
These, as successive generations bloom,
New powers acquire and larger limbs assume;
Whence countless groups of vegetation spring,
And breathing realms of fin and feet and wing.

Thus the tall Oak, the giant of the wood
Which bears Britannia's thunders on the flood;
The whale, unmeasured monster of the main,
The lordly lion, monarch of the plain,
The eagle soaring in the realms of air,
Whose eye undazzled drinks the solar glare
Imperious man, who rules the bestial crowd,
Of language, reason, and reflection proud.
With brow erect who scorns this earthly sod,
And styles himself the image of his God;
Arose from rudiments of form and sense
An embryon point, or microscopic ens!

Darwin, 'The Temple of Nature' (1803)

What a wonderful, evocative – if rather baroque – summary of evolution this is! It's surprisingly accurate, too, with its description of life beginning on a sub-microscopic scale in the oceans, and a hint that whales appeared long after life had moved to the land. It even culminates with the assertion that Man himself is part of this great chain of life, the assertion that is said to have so outraged devout mid-Victorians. Yet the Darwin who wrote this was not Charles, the author of *On the Origin of Species*, but his grandfather Erasmus, and it was written almost 60 years earlier than the famous book. To those brought up to believe that Charles Darwin unleashed the idea of 'evolution', especially human evolution, on a shocked and unsuspecting public, this is perhaps a bit of a surprise.

In fact, the idea that species might change over time is a very old one, dating back to the days of ancient Greece. Over 2,500 years ago, Anaximander (c.610–546BC) reportedly suggested that life began in the sea and only later emerged on land. And consider this description by Aristotle of the ideas of Empedocles – a thinker who lived in the fifth century BC – on how life came about: 'Wherever then all the parts came about just what they would have been if they had come to an end, such things survived, being organised spontaneously in a fitting way; whereas those which grew otherwise perished and continue to perish.' That sounds surprisingly similar to natural selection and survival of the fittest, doesn't it?

Listen, too, to this description from the Islamic thinker al-Jahiz, writing in ninth-century Baghdad: 'Animals engage in a struggle for existence [and] for resources, to avoid being eaten and to breed ... Environmental factors influence organisms to develop new characteristics to ensure survival, thus transforming into new species. Animals that survive to breed can pass on their successful characteristics to [their] offspring.' That appears even closer to Darwinism. Of course, the translation of this fragment of text here gives it an especially modern spin, and I'm not suggesting in any way that Empedocles or al-Jahiz beat Darwin to it. But the point is clear. Not just hundreds but thousands of years before Darwin, acute thinkers were speculating on just how life could have developed on earth without resorting to supernatural magic, and coming up with ideas that contain at least the germ of evolution.

Evolving ideas

So the notion that when Darwin (Charles!) wrote about 'evolution' it was a thoroughly shocking and entirely new idea is complete nonsense. This is not just an arcane matter of historical misunderstanding; it matters because in embracing Darwin as a heroic individual pioneer, and the idea of 'Darwinism' or 'neo-Darwinism' rather than evolutionary theory, we are perhaps creating an intellectual straitjacket for ourselves. It may also perpetuate the myth that science remains static for a long time, then suddenly makes a giant leap forward through remarkable flashes of insight of revolutionary brilliance from a great thinkers like Darwin. This also does a disservice to Darwin who was always eager to engage with others' opinions and ideas.

The way that 'science' develops over time is much more complex that it might first appear. The historical construction of icons of science can sometimes cloud the bigger picture. It also makes the individuals concerned seem less human, less fallible, which in turn provides an opportunity for pointless pot shots about how they were 'mistaken' or 'wrong' about different facets of their work. Darwin is worth celebrating and is clearly of very great importance in the history of evolutionary theory, but he was part of a wider tradition of people who were interested in whether species might change over time. And although the main core of Darwin's theory – evolution by means of natural selection – does form the basis of most modern evolutionary biology, with the benefit of scientific hindsight Darwin was not 'right' about everything, nor should we expect him to have been. What I would like to do is put Darwin's breakthrough in context, not just in order to set the record straight but because it is a fascinating story in itself.

From the Greeks to Darwin

The science of natural history has a long tradition with its roots in antiquity. Ancient thinkers were interested in observing and describing all aspects of the natural world including minerals, animals and plants alongside astrological and meteorological phenomena. The works of Aristotle (384–322BC) on zoology and those of his pupil Theophrastus (c.381–276BC) on botany formed the basis of natural historical writing right up until the mid-fifteenth century, along with the work of others such as the Greek physician Dioscorides (c.AD40–90) and the later Roman encylo-paedist Pliny the Elder (AD23–79).

However, natural history as an academic pursuit in Western Europe languished from the first century until the Renaissance. The naturalist responsible for setting up the first botanical garden and public museum in Bologna, Ulisse Aldrovandi (1522–1605), noted that the study of natural history was a 'faculty buried for so many years in the gloom of ignorance and silence'. However, even though the subject enjoyed a remarkable revival during this period, the natural history of the Renaissance still maintained the encyclopaedic framework of the much earlier work of Pliny and at times still focused on the emblematic or mythological impor-tance of certain animals. Perhaps the most widely read of all of the Renaissance works of natural history was *Historia Animalium* (*History of Animals*), written by Conrad Gesner (1516–65). The volumes of this natural history listed a wide range of animal species, focusing on different categories such as viviparous quadrupeds (four-footed animals that bear living young) or oviparous quadrupeds (animals that hatch their young from eggs). The natural history of the Renaissance, though in some ways concerned with how to categorise species, was vastly different to natural history today – it did not employ the same level of analysis as the later systematic descriptions of species that would truly enable discussions about both the diversity of species and the origin of those species.

During the seventeenth century, in a period that is often referred to as the 'scientific revolution', a number of ideas began to engender the start of a profound shift in the worldview of intellectuals and lay people alike. This is an age typified by the well-known icons of science: Galileo, Newton, Boyle and Hooke. Natural phenomena from the movement of the universe to the microscopic world began to open up for further investigation. As Europe moved through the Enlightenment, countless thinkers began to make their own observations and deductions about the world. Some of them are relatively well known, such as Priestley and Lavoisier who battled over the nature of atmospheric gases and began to lay the foundations of modern chemistry, and William Herschel who first discovered Uranus. But there were many, many others, now long forgotten, who made their own small contributions to the rising tide of ideas and discoveries, observing, exploring, experimenting, writing, thinking or simply just discussing ideas. The advent of the Industrial Revolution, with its new machines and inventions reaching even into lowly spinners' cottages, reinforced the general belief that things work in a mechanical, cause-and-effect, predictable way.

The exuberant planet

Throughout the seventeenth, eighteenth and early nineteenth centuries, botanists and zoologists were learning to appreciate the incredible diversity of natural life in the world. Some studied the plants and animals around them; others looked at exotic plants and animals brought back from the distant places that ships were now visiting. Collecting plants became almost an obsession – especially in England, where Kew Gardens was established in 1759 – and the gardens of not just stately homes but even modest vicarages and villas began to fill up with a cornucopia of plants from every corner of the world. Sir Joseph Banks, who had made his own fantastic collection of plants on Cook's voyage to Australia, became the unofficial director of Kew Gardens from 1773 and organised armies of collectors to scour the world for exotic plants to add to Kew's collection.

Darwin's *Beagle* voyage of the 1830s was a latecomer, but very much part of this almost greedy fascination with the teeming variety of life around the world. Today botanists and zoologists are a little more circumspect in their specimen-hunting, but from the early eighteenth century onwards flowers and trees, insects and birds were all collected, described and illustrated, dug up and potted, caught and pinned, shot and stuffed, boxed and carted back to Europe from all over the world.

The sheer diversity of natural life revealed by these nature hunters during this period was staggering, and even professional botanists and zoologists struggled to make sense of it all. It was a source of wonder, as the exuberant poems of Erasmus Darwin and many others show. But it was also a source of much activity as they tried to put it into some kind of order.

Cataloguing Creation

One of the first great attempts to find some pattern in the profusion came from the English botanist John Ray (1627–1705), who devised a new scheme for classifying plants. Ray's plant classification system was based on the seeds produced by flowering plants. He was the first to make the distinction between plants that are monocotyledons (those with one leaf on the seed) and dicotyledons (which have two leaves on each seed). Sir Hans Sloane the eminent collector of all things including plants* was one of Ray's correspondents, and it was these correspondents who helped him to describe a total of 16,100 plant species in his three-volume *Historia plantarum* (1686–1704).

Ray was an influential natural theologian – perhaps his most famous work is not his important contributions towards botany but *The Wisdom of God Manifested in the Works of the Creation*, published in 1691. This was a very influential work in which he reconciled his extensive botanical knowledge with his theism. The study of the natural world, he argued, suggested the role of a benevolent God in its design.

Half a century later, the botanist Carl Linnaeus (1707–78), sometimes known evocatively in his native Sweden as the 'Flower King', took up where Ray left off to create his definitive *System of Nature*. His work on plants, *Species plantarum* (1753), and the tenth edition of his book on animals *Systema naturae* (1758), are

---

* Sir Hans Sloane's collection, which consisted of 71,000 objects and 50,000 books including his herbarium, eventually ended up forming the basis for the collections of the British Museum after his death. Sloane also has the honour of having introduced and promoted milk chocolate to the UK – thank you Hans!

seen as the starting points for modern botanical and zoological taxonomy.

Linnaeus was not alone in trying to find a classification system. By 1799, a significant number of schemes had been proposed. But Linnaeus's system had two key features which guaranteed its survival. First of all, Linnaeus grouped plants and animals according to a carefully chosen range of anatomical features; for instance, he grouped plants by their flowering parts and realised that mammals could be identified and unified into a single group by their possession of mammary glands. Second, he gave each species a consistent two-part Latin name, such as *Passer domesticus* (house sparrow), the first part of which is always the name of the group or genera it belongs to and the second of which is the species name. This system was so powerful and effective, as was its promotion by Linnaeus' pupils, that it was adopted by botanists and zoologists around the world by the end of the eighteenth century and has remained in place ever since. The Linnaean system was fairly rapidly adopted in Britain; for instance Sir Joseph Banks was the patron of one of Linnaeus' pupils David Solander. They went on a number of expeditions together, including one under Captain Cook on the *Endeavour* to Tahiti. Solander eventually became the curator of Banks' natural history collection and in 1773 became keeper of the natural history collections in the British Museum. The acceptance of Linnaeus' system in Britain was such that just ten years after Linnaeus' death in 1778, the Linnaean Society was founded in London.

The neat classification of life in Linnaeus's system, with its division of all the earth's organisms into groups and subgroups, gave biologists for the first time a coherent framework within which to view the teeming variety of the natural world.

The ancient earth

An understanding of the age of the earth is as central to the explanation of evolution as an understanding of the relationships between the many diverse species. Most people are under the impression that before the nineteenth century and Darwin there was a very truncated view of the history of the earth. The example that is often cited is the work of the seventeenth-century prelate James Ussher (1581–1686), who calculated that the world began precisely on Sunday 23 October 4004BC and suggested that little had changed since, except perhaps during the time of the deluge, the biblical flood, which he dated to 2349BC. While this may seem ridiculous to us today, we have to at least concede that this was a phenomenal piece of biblical scholarship!

In the seventeenth and early eighteenth centuries, thinkers like Thomas Burnet (c.1635–1715) and his avid supporter Isaac Newton (1643–1727) studied the Bible and other historical sources to learn about the history of the earth. However, there were those – such as Gottfried W. Leibniz (1646–1716) and Benoît De Maillet (1656–1738) – who had begun to realise that the formation of the modern world could not have taken place through natural processes within the time stated in the literalist Genesis accounts of creation. The evidence from observation simply supported the idea that the earth was much older.

Others, like Nicolas Steno (1638–86) and Robert Hooke (1635–1703), began to argue that fossils were the remains of living organisms. Steno realised how the relative depth of layers of rock indicated their age, and quite soon it became clear to many that the world was really very ancient – and hid its own long and fascinating story. Keen observers began to pay closer attention to fossils, and began to see that they hinted at a much richer and bigger story about the history of the earth and the life on it than most people had ever imagined. As naturalists began to build fossil collections, they realised that some fossils were remnants of creatures no longer to be found on earth.

Towards the end of the seventeenth century, the interpretation of the Bible's creation story which implied that the earth had been formed quite recently with life ready-made was, for some, already looking decidedly more and more like metaphor than literal truth.

As old as the hills

At the same time, following on from Steno's work, geologists began to take an interest in the implications of fossils and just how the rocks in which they were found had formed. Geologists had noticed that many rocks, even in high mountain ranges, were clearly formed from sea-floor sediments, and that they were full of the fossils of sea creatures. Yet how did the sediment from the ocean bed come to form mountains – and how did the fossils get there?

A number of theories were being suggested as to how the earth had formed. One of these became known as Neptunism. The great German geologist Abraham Gottlieb Werner (1749–1817) was Neptunism's leading proponent. Most rocks, he suggested, formed in a universal ocean that covered the entire earth, and were then left behind as the landscapes we see today when the ocean waters drained away. That's how fossils of sea creatures got on top of mountains. Of course, some saw the biblical deluge in Werner's universal ocean – though this was not necessarily a majority viewpoint. And later geologists suggested a series of catastrophes of which the biblical deluge was just the most recent.

Others suggested a set of theories that became known as Vulcanism. The surface of the earth was, they argued, the result of prolonged volcanic action and earthquakes. This view was extended by the mild-mannered Scottish geologist James Hutton (1726–97). Hutton could find no evidence of a single catastrophic, universal flood. But he could find all kinds of local evidence that landscapes had been shaped not once and for all in some great catastrophe, but slowly and continually by countless cycles of erosion, sedimentation and uplift – repeated over incredibly long time periods. If this is so, then the earth must be very, very old – not just thousands of years but millions. Hutton contested that this was the result of a benevolent God who had designed the natural processes of a world free of catastrophic actions that would thus be fit for human habitation in perpetuity.

Rock of Ages

Soon afterwards, the English surveyor William Smith began to make acute observations of the rocks as he surveyed routes for the new canals that were etching their way across the new English landscape of the Industrial Revolution. He was probably not the only surveyor who made observations like this, but he is the one we know about. An acute and diligent practical observer, Smith found that each rock layer contained its own distinct range of fossil types. Then he realised that rocks that outcropped long distances apart, but which contained the same range of fossils, must have formed at the same time. So using these fossil assemblies, he could identify rock outcrops and establish their relative age right across the landscape. In this way, Smith created the first geological map in 1801, and in 1815 published the first geological map of Great Britain.

The implications of Smith's work were huge. Fossils occur in sedimentary rocks; that is, rocks that have been formed mainly by the age-long compression of sediments that have been laid down

on the floors of oceans and lakes. The numerous layers identified by Smith showed that Hutton was right. It was pretty conclusive evidence that there really had not been just one almighty
catastrophic event in the earth's history at all. Instead, each rock
layer revealed one of many episodes of erosion and uplift that had
been going on for a very long time. Studying the different layers
of rocks and their characteristic fossils was like opening a marvellous, giant book in which the earth's long and convoluted history
was unfolded, age by age.

The work of Smith, Hutton and numerous other British geologists finally coalesced and was introduced to a wider public in
1830 in Charles Lyell's landmark book *The Principles of Geology*,
which is for geologists as much of a turning point as Darwin's
On the Origin of Species is for biologists. On the first page of his
book, Lyell provided a wonderfully graphic illustration to demonstrate the core idea of the book – that the earth has been shaped
and reshaped again and again over vast periods of geological time
by repeated cycles of gradual erosion and uplift. The frontispiece
illustration was an engraving of the ancient temple of Serapis
from Pozzuoli on the Italian coast. High up on the temple col-
umns are countless dark holes bored into the stone by shellfish
when the temple was submerged beneath the sea long ago, before
it re-emerged to reveal the holes later on – striking evidence of
Hutton's contention that whole blocks of land can move up and
down over time.

The young Darwin read Lyell's *Principles of Geology* whilst
on his famous round-the-world voyage on HMS *Beagle*. It was
profoundly influential on him and it is quite clear that Darwin's
vision of the slow and gradual evolution of life, carrying on
over vast periods of time, owes much to Lyell and his picture
of slow geological processes operating over unimaginably long

eras. Indeed, Darwin clearly acknowledged his debt to Lyell who became a close friend. Lyell's uniformitarian geological history of the earth became the arena in which Darwinian evolution could be played out.*

'By far the best proof is experience'

There are a few more strands I would like to mention briefly before returning to Darwin's immediate milieu – particularly the development of philosophical, political and economic ideas in the eighteenth century. A number of interesting intellectual movements underpin our understanding of what science is today, many dating back to the close of the Renaissance.

Key in England, at least, was the work of Sir Francis Bacon in the early seventeenth century. Bacon wanted to overthrow the shackles of Aristotelian logic and its insistence that knowledge could be arrived at through deduction and induction. Deduction is a process whereby a set of assumed premises or axioms are used to infer a theory about the world. It is a process based in this context on the use of logic, not observation. For Bacon, real knowledge of the world could only be achieved inductively. Put another way, he thought that in order to learn about the natural world we have to free our minds of assumptions, make observations and experiments involving natural phenomena, and look objectively for patterns that then lead to axioms and theories.

---

* Lyell himself struggled to accept evolutionary theory. Although later in his career, in the tenth edition of *Principles of Geology* (published 1867–8), he supported the idea that evolution was responsible for the origin of new species and that man was even a part of this process, he never allowed that Darwin's theory could adequately account for the mental and moral aspects of the human species.

This could never be a solitary activity, but depended on the exchange of ideas and observations. Bacon's inductive method laid the foundations for the advancement of science in the seventeenth and eighteenth centuries – and his vision of Solomon's House, a scientific research institute described in his utopian novel *The New Atlantis* where the best minds would exchange observations, was the inspiration for the influential scientific societies that came to dominate science in Britain and France. The Royal Society in London, for instance, was set up in 1660 by followers of Francis Bacon. This way of looking at things gave a crucial intellectual stimulus to the great scientific discoveries of the seventeenth and eighteenth centuries and the inquisitiveness about the natural world that led to the explanation of evolution – and it also laid the groundwork for the way science is practised today, as we will see later.

Reason or experience?

Baconian emphasis on induction instead of deduction also fed into the perceived split between 'empiricism' and 'rationalism'. At the heart of the split lay a question which has been a preoccupation for the entire history of Western philosophy: how do we come to know things? Do we acquire knowledge mainly through our use of reason? Or do we acquire it through experience gained with our senses? This dichotomy between reason and experience, rationalism and empiricism, has to some extent been at the heart of intellectual enquiry in the West ever since the seventeenth century.

Put simply, the core of this debate is that rationalists argue that our knowledge is gained independently from what we experience through our senses. For instance, the French mathematician René Descartes (1596–1650) wanted to discard any belief that is open to the slightest degree of doubt – arriving at his famous bottom line *cogito ergo sum*, 'I think therefore I am'. This is the ultimate rejection of the senses as the sources of certain knowledge. Only the mind, conscious of its own existence by the act of thought, could not be deemed fallible.

Rationalists might argue that reason or logical thinking has precedence over all other ways of acquiring knowledge. Empiricists, on the other hand, argue that any attempt to acquire knowledge should be based on sense experience – namely, observation of the natural world. Therefore, the path to knowledge is through observation and experimentation. It is for this reason that empiricism is seen as the basis of modern life sciences.* We will return to this debate in more detail in chapter 7.

The religion of nature

You may remember from the introduction to this book Dawkins' version of the role of the early-nineteenth-century theologian William Paley (1743–1805). Dawkins cites Paley's assertion, that an object as complex as a watch must be the creation of a watchmaker, as a perfect example of the religious delusion that Darwin finally saw through. Well, any philosopher will tell you that the Scottish empiricist David Hume dealt with this argument pretty effectively almost 80 years earlier than Darwin. In his *Dialogues*

---

* We should be a little wary of categorising individuals into two defined camps – the thought processes of individuals are much more subtle than the grandiose divisions that have been retrospectively applied. It is difficult to sort different thinkers of the early modern period into the categories 'empiricist' or 'rationalist', in part because they picked out different phenomena as objects of 'experience' or embraced different aspects of the theories concerned. The usual suspects in this dubious split are the Continental rationalists Descartes, Leibniz and Spinoza versus the British empiricists Locke, Berkeley and Hume – and of course it is never really that simple!

Concerning Natural Religion, published posthumously in 1779, Hume beautifully uses one of his characters, Cleanthes, to argue that the amazing intricacy and complexity of the world around him just shows that it must have been designed by a mind with human or more than human intelligence, thereby proving the existence of a benevolent God.* Then he goes on to demolish Cleanthes' argument entirely through the arguments of a character called Philo.†

We shall return later to this argument from design, which is of course exploited by modern-day creationists, but the point is that Paley's position was far more nuanced than Dawkins implies. Paley was not simply the old-fashioned religious antecedent to Darwin's radically new scientific insights. In fact, Paley wrote partly in response to Hume, arguing that one could not simply ignore one's first perception that nature showed design, even if it was possible to slowly demolish the perception later through Philo-like logical argument.

The subject and title of the book by Paley to which Dawkins was referring was *Natural Theology* (1802). 'Natural theology' is the notion that God can be known as well through nature as through supernatural revelation. It was not a new idea, dating back to the time of St Paul, but it became increasingly popular in the eighteenth and nineteenth centuries. In some ways it allowed a more enlightened, common-sense view of the world than the

---

* The argument that apparent design shows there is a God is known as the 'teleological' argument, from the Greek *telos* which means 'end' or 'purpose', since the appearance of design in nature is thought to show that there is a purposeful agent behind it.

† Immanuel Kant, too, criticised the argument from design in his attempt to steer a line between the rationalists and the empiricists, but his arguments were more subtle than Hume's, criticising not the idea of design per se, but the unjustified assumption of purpose in design.

straitjacket of revelation required and, more significantly, had no necessary points of conflict with the scientific discoveries of the day.

Divine nature

Natural theology became more prominent in the first half of the nineteenth century in response to the changing landscape of knowledge about the natural world. The marvellous diversity in nature and the almost miraculous appearance of design were thought to be, as Paley suggested, evidence of God's hand at work. Some natural theologians argued that God's work was shown in the sheer cleverness and appropriateness of so many natural adaptations, as Paley argued with his picture of the divine watchmaker. This particular argument reached its peak in the famous series of specially commissioned works called the *Bridgewater Treatises* of the 1830s. Others insisted that God's transcendent power was shown through underlying similarities in nature that could be discovered amid the diversity. The first director of London's Natural History Museum, Richard Owen, developed a theory of archetypes – divine plans – which he searched for through meticulous comparative anatomy.

Unlike Darwinian evolution these varying natural theological arguments still insisted on a definite role for God, but they progressed and altered in response to new discoveries, and it was quite possible for a natural theologian to be a leading scientist like Owen, working at the cutting edge of science. Indeed, natural theology played a key role in enabling the discoveries that led to the development of Darwinian evolutionary theory. The countless amateur naturalists and fossil hunters who added to the growing store of scientific knowledge could indulge in their favourite hobby with the spur and sense of worthiness that came from knowing they were looking for signs of God's handiwork. Darwin himself read Paley's *Natural Theology* when he was at Cambridge. Although Darwin dismissed Paley's argument from design in his *Autobiography*, he later said of *Natural Theology* and other works by Paley that they 'gave me as much delight as did Euclid. The careful study of these works, without attempting to learn any part by rote, was the only part of the academical course which, as I then felt and as I still believe, was of the least use to me in the education of my mind.'

God's mechanics

However William Paley, some believed at the time, was a closet deist, because of the persistence of mechanistic images of nature in his books. By Paley's time, deism was weakening in its influence, but through much of the seventeenth and eighteenth centuries it had and continued in some forms to be part of a powerful line of radical religious thought, espoused by such great minds as Newton and Voltaire. Indeed the founders of the American constitution, notably Thomas Jefferson, James Madison and George Washington, are all thought to have been deists. So too was Thomas Paine.

Deism was, in some ways, a way of reconciling belief in God with the rational pursuit of knowledge. Deists believe that there is no need for mystical revelations and the proof of miracles to arrive at the truth; human reason alone is enough, they argue, because God has created the human mind with the capacity to reason, and created the universe in such a way that it can be understood by reason. In some ways, the universe is seen simply as a machine created and set in motion by God.* That was going

---

* This might seem at first sight a little like modern-day creationists' idea of Intelligent Design. In fact, it is fundamentally different, since it espouses a view

much further than natural theology which simply said that God is revealed through nature, and is also the reason why the mechanistic images in Paley's work hint at deism. Deism was taken up by radicals in the eighteenth century partly because of the prominence it gave to human reason, which was within the reach of every single man and woman, as opposed to supernatural revelation which might only be granted to the select few.

Deist ideas became deeply woven into the thread of the French Revolution and were espoused by Robespierre among others, and no doubt its associations with the horrors of that time helped to give it a bad name in Britain. Intriguingly, the theory of the transmutation of species also got a bad name in England because of its association with French radicalism. By that time, though, deism was already under attack from many directions. Hume and many others undermined its basic logic. In France, many materialist thinkers went the logical step further to eliminate God even from the creation of the universe.

So when Paley wrote *Natural Theology,* it was far from a simple expression of old-fashioned creationism just waiting for Darwin to come along and correct it. He was actually responding to a wave of current ideas on transmutation (which we shall explore shortly) in a way that belonged to a tradition of thinking endorsed by Newton and Priestley, Franklin and Voltaire, Montesquieu and Jefferson. His book was actually quite successful, selling 15,000 copies – not perhaps because it was reassuring old religious

_____

of the universe that works entirely and completely by natural, mechanical laws and processes, even though these laws are said to have been divinely created. This means that a deist would have absolutely no problem in accepting an entirely scientific theory of evolution, unlike 'Intelligent Design' adherents.

dogma but because it actually dealt with contemporary ideas in science and philosophy.*

Interestingly, for some, the French Revolutionary taint to deism almost meant that the inference that Paley was a deist could be taken as a slur, hinting that he was a bit of a continentalist radical. In fact, though Paley was far from radical politically, he was an ardent advocate of the abolition of the slave trade, a latitudinarian whose unorthodox views barred him from high office and, above all, a highly respected philosopher whose lucid political writings were standard texts at Cambridge for almost a century.

Free thinking

Deist thinking probably influenced another of the eighteenth century's most influential thinkers, the Scottish economist and moral philosopher Adam Smith. Smith believed, like the deists, that the world was controlled entirely by natural laws, rather than by supernatural events – a great machine, designed by God, but in which he did not interfere. Of course, Adam Smith is most famous for his hugely influential economic theories, expressed principally in *The Wealth of Nations*. Smith believed that economies are continually changing and developing, driven by the way in which every individual acts in their own self-interest – but the combined effect of all these individual activities works for the wealth and benefit of all. In *The Wealth of Nations* he argued that 'Every individual is continually exerting himself to find out the most advantageous employment for whatever capital he can

---

* It is interesting that the poet Shelley, in the Preface to *Prometheus Unbound*, brackets Paley with Malthus, one of the inspirations for Darwinian natural selection, as figures who reduce the human spirit: 'For my part I would rather be damned with Plato and Lord Bacon, than go to heaven with Paley and Malthus.'

command. It is his own advantage, indeed, and not that of the society, which he has in view. But the study of his own advantage naturally, or rather necessarily, leads him to prefer that employment which is most advantageous to the society.'

This idea of an evolving economy driven by individual needs may well have played a part in the development of evolutionary ideas and the theory of natural selection. There is a clear similarity between Darwin's theory of evolution and the rise and fall of corporations, with some surviving and the weaker and less adaptable falling by the wayside. We even know that Darwin read Adam Smith. However, I am not trying to assert a direct connection at all. What I am suggesting is that Adam Smith's ideas were part of a changing way of looking at the world that paved the way for Darwin's ideas – a realisation that the complex, teeming, superficially random variety of the world could move and develop in tune with natural laws that could be discovered and gradually understood. Of course there *is* an often-cited connection between Darwin and another great economic theorist of the later eighteenth century, Thomas Malthus, as we will see in the next chapter.

Yet as Darwin wrote in his *Autobiography*, he was already familiar with the continual struggle for existence from observing nature, and the poet Tennyson famously wrote of 'nature red in tooth and claw' long before Darwin published *On the Origin of Species*. Indeed, the whole Romantic movement that reached its peak in Darwin's formative years was driven by the forlornness of the individual in a difficult world. The Romantics emphasised their alienation from the increasingly competitive and hectic urban and industrial world. But the wildness of nature, though inspiring, had a hostile face too, as countless tempest-tossed visions testified.

To summarise, the way that the natural world works, the nature and number of species and how they might change over time is a line of thought that has its roots in antiquity. It is simply wrong-headed to imply that there was a sudden shift in thinking from creationism to evolution after Darwin published his theories. There simply was not one set of concrete ideas – such as 'creationism' – that Darwinism replaced, as Dawkins has sometimes implied. In fact, what we tend to view as creationism today is a historical construct – it is a by-product of both the way in which the science and the promotion of Darwinian evolutionary theory has itself evolved, and how that historical process has been described.

What we understand by Darwinism today is also in some ways a historical construct – Darwin's theories and Darwin himself have been subject to so much attention particularly in the past 30 years that we perhaps tend to have a rather narrow, iconic image of the way in which evolutionary theory developed. Perhaps surprisingly, though, Darwinism has not always enjoyed such a central place in the historical analysis or indeed celebration of evolutionary theory. And ironically there are plenty who have tried to name other 'precursors' to Darwin as the founders of evolutionary theory. We will in part explore why this has been the case in the following chapters.

CHAPTER 2

DARWIN PRECURSED

Ignorance more frequently begets confidence than does knowledge: it is those who know little, and not those who know much, who so positively assert that this or that problem will never be solved by science.

Charles Darwin, introduction to
The Descent of Man (1871)

Spotting precursors of Darwin has become something of a cottage industry. There have always been those who wish to prove themselves in some way, or prove a point, by claiming the mantle of evolution for some other neglected genius who saw it all first. There are other revisionists who want to rewrite history altogether from another perspective. I have no wish to do either, because I believe Darwin rightly has pride and prominence of place as the father of modern evolutionary theory. It was Darwin, and only Darwin, who had the vision, persistence and clarity of thought to turn it from a great idea into a comprehensive theory that convinced others and has more than stood the test of time. However, I would like to introduce some of these so-called precursors because of the way in which the story has influenced modern-day Darwinism.

It is this precursor-hunting, coupled with the equally dangerous portrayal of Darwin as a lone voice in the wilderness or an iconic revolutionary figure, that most often leads to a misrepresentation of his context, theories and ongoing legacy in evolutionary theory today. Darwin had two particular direct inspirations for his work – Charles Lyell's *Principles of Geology* and Thomas Malthus' *Essay on the Principle of Population*.

In October 1838, Darwin read Malthus' *Essay on the Principle of Population* and was apparently hugely struck by his assertion that unchecked population growth would eventually outstrip food supply so that in time the population would be limited in the most brutal way. Darwin recognised that this meant that not all those born would survive, and that it was the most vulnerable who would die first. The parallels between Malthus' stark view of population limits and Darwin's view of the struggle for existence are unmistakable, and this recognition clearly played a role in helping him to crystallise his theory of natural selection. Charles Lyell, who had perhaps a more lasting impact on Darwin's thinking, was an influential geologist whom I will discuss in more detail below.

Darwin insists that he was not influenced by any other evolutionary theory – especially not that of the now infamous Lamarck. He wrote in his *Autobiography*:

> It has sometimes been said that the success of 'The Origin' proved 'that the subject was in the air' or 'that men's minds were prepared for it.' I do not think this is strictly true, for I occasionally sounded out not a few naturalists, and never happened to come across a single one who seemed to doubt the permanence of species.

While this may have been the case in the circles within which Darwin was discussing his ideas, and was almost certainly true when it came to the concept of evolution by means of natural selection, there was already a fine tradition across Europe of challenging the permanence of species – both among men of science and in more public spheres. Some of these earlier ideas have impacted heavily on the way that Darwinism, or more importantly neo-Darwinism, is discussed today. Again, it is important to understand the historical context of Darwin's theory in order to comprehend Dawkins' role in promoting his own version of Darwinism.

Buffon's heirs

In France and Germany the idea that species transmute, or transform, over time had begun to emerge as a coherent idea, and some continental historians quite rightly suggest that the focus on Darwin alone is a very British obsession. In fact, one of the first great pioneers of what we now think of as evolutionary thought was the Frenchman Georges-Louis Leclerc, Comte de Buffon (1707–88). Buffon was an extraordinary figure, dazzlingly brilliant, literate and insightful – and apparently a magnet for women! Jean le Rond D'Alembert, the mathematican and philosopher who was critical of aspects of Buffon's work, called him 'the great phrasemonger' for his remarkable ability to coin a telling phrase. He was something of a polymath, delving into mathematics and all branches of science, but it was in natural history that he really made his name. The evolutionary scientist Ernst Mayr said, glowingly: 'Truly, Buffon was the father of all thought in natural history in the second half of the 18th century.'

It was Buffon, among others, who began the business of scientifically pushing back the age of the earth. It had long been known that the earth radiated heat, as the warmth of deep mines and the heat of volcanoes showed. In order to estimate the age of the earth, Buffon conducted an experiment in which he heated solid iron globes and measured how fast they cooled down. In his first volume of *Histoire naturelle*, published in 1749, he estimated the age of the Earth to be at least seventy thousand years. He apparently attracted censure from the Church, so he retracted his estimate, and only published it again 30 years later when theories that suggested a longer earth history were becoming more widely accepted. Buffon was one of those who had started to nudge ajar the door to the idea of an ancient earth, and all that remained was to push it fully open, as Hutton and other geologists started to do.

Buffon was appointed superintendent of the Jardin du Roi in Paris in 1739, and he oversaw its transformation from a royal physic garden to a leading research centre for natural history. He was therefore recognised as France's leading naturalist. He is most well known for his stinging criticism of Linnaeus' system of classification, but a more important insight was his conception of what a species is – it is in these criticisms that some see the seeds of Darwinian evolution.

Darwin himself said that 'the first author who in modern times has treated it [evolution] in a scientific spirit was Buffon ...', and once remarked to T.H. Huxley: 'I have read Buffon; whole pages are laughably like mine.'

Buffon's works, though, are so voluminous that it is understandably difficult to establish just what his ideas were. However, scattered through Buffon's work are some of the key elements of what we understand today as evolutionary theory: the idea that organisms should be understood as individuals within a species; and the idea that species diverge from a common ancestor. But we must remember that Buffon was working within a completely different conceptual framework from that of Darwin. For example, he hypothesised that there was an underlying pattern to the forms of life that were possible. Buffon also subscribed to the materialistic belief of the time, namely spontaneous generation, and argued that the original, ancestor forms of families may have naturally spontaneously generated due to exceptional environmental conditions. He thought that there was some pattern to this spontaneous generation. He did not, however, believe that this pattern was organised in any way by a divine creator or that it adhered to a divine design. Rather, Buffon suggested that the very nature of the universe predicted and limited the form of organic beings (much as, as we now know, chemicals can only form certain stable isotopes), so that given the same conditions on other planets similar life forms would occur. Darwin suggested that where he and Buffon differed was that unlike him, Buffon 'does not enter on the causes or means of transformation of species'.

Out of the Terror

Towards the end of the eighteenth century, transformist ideas* began to bubble up again and again in France and Germany despite, or maybe because of, the turmoil of the French Revolution. Right in the heart of Revolutionary Paris in June 1793, a new museum of natural history was founded, the *Muséum National d'Histoire Naturelle*. The man who became its first professor of vertebrate zoology was Étienne Geoffroy Saint-Hilaire (1772–1844). The museum's first professor of invertebrate zoology – and the man who coined the term 'invertebrate' – was the most famous by far of Darwin's predecessors, Jean-Baptiste-Pierre-Antoine de Monet, Chevalier de Lamarck (1744–1829).

Geoffroy Saint-Hilaire was just 21 at the time, but he proved remarkably energetic and effective. A short while after his appointment, he brought another young naturalist, George Cuvier (1769–1832), to the Institute. Between them Geoffroy Saint-Hilaire,

---

* Transformism, loosely speaking, suggests that current species have gradually transformed, or transmutated, from other parent species.

Cuvier and the older Lamarck effectively established the sciences of biology and palaeontology, laying much of the basic groundwork for Darwin's theory (in Cuvier's case inadvertently).

Form vs. function

Geoffroy Saint-Hilaire is perhaps the least well-known of the three, but he established the crucial concept of homology – looking for correspondences of characteristics between different organisms. Geoffroy Saint-Hilaire, a deist argued that there was an underlying unity between divergent forms and though this did not necessarily imply a form of common descent we might recognise today, this was the essence of his transmutational ideas. Geoffroy Saint-Hilaire proposed that new species were formed suddenly due to changes in the environment, though as we shall see he did not develop or publicise his theories in the same way that Lamarck did his. Identifying homologies was to become a crucial way of teasing out evolutionary relationships between creatures that looked superficially very different, and remains a valuable but sometimes controversial approach today.

Darwin was keenly aware of how important Geoffroy Saint-Hilaire's work was, writing, in the first edition of *On the Origin of Species*:

> What can be more curious than that the hand of a man, formed for grasping, that of a mole for digging, the leg of the horse, the paddle of the porpoise, and the wing of the bat, should all be constructed on the same pattern, and should include the same bones, in the same relative positions? Geoffroy St. Hilaire has insisted strongly on the high importance of relative connexion in homologous organs: the parts may change to almost any extent in form and size,

and yet they always remain connected together in the same order. We never find, for instance, the bones of the arm and forearm, or of the thigh and leg, transposed. Hence the same names can be given to the homologous bones in widely different animals.

Cuvier, on the other hand, was a great believer in function rather than form. He insisted that if there are any resemblances between the body parts of different organisms, it only matters when they are performing the same task. Resemblances of form alone, he contended, are essentially imaginary. The rivalry between Geoffroy Saint-Hilaire and Cuvier, and these two diametrically opposed ideas, came to a head in a series of key debates in 1830. Cuvier, a meticulous anatomist, came out on top, showing that many of Geoffroy Saint-Hilaire's more extreme homologies were simply not plausible, but the form vs. function debate in some ways remains a hot one in evolutionary biology even today.

The fossil king

As the foremost animal anatomist of his day, Cuvier was renowned for his ability to identify a creature from just a small fragment of bone. He was the first to show that some fossils are from animal species that are now extinct. When elephant-like fossils were discovered in Siberia and America using his brilliant anatomical analysis skills, Cuvier was able to show that they belonged to extinct species which would become known as the mammoth and the mastodon. He then went on to show the existence in the distant past of many other large creatures that had long since vanished, such as the Irish elk and the giant ground sloth. However, Cuvier was vehemently anti-transformism (or evolution as we might call it today) and adhered to the idea that species could not change over time. He argued that the structures found in species that he had studied were so finely tuned that they could not be changed without causing considerable disruption to the function of the organism. According to Cuvier, the extinction of species shown by the fossil record was not caused by species changing over time from common ancestors. In his view, species did not change; they just died out. The Napoleonic expeditions in Egypt seemingly supported Cuvier's viewpoint – the mummified species sent back were clearly thousands of years old but were no different from species in existence today.

Cuvier did not necessarily accept that species that replaced those that became extinct were created by any divine intervention. Rather, he argued that extinction was caused by catastrophic events, each of which was restricted to one particular region or continent. Thus species outside these stricken regions simply moved into the areas vacated by the extinct species. According to this argument, species that had appeared in the fossil record in Europe in later periods had simply moved there from other continents. The increasingly pressing problem with Cuvier's theory was that as exploration allowed for the global fossil record to be studied, it became clear that there was no evidence of the existing European species in the ancient fossil records of other continents or vice versa.

Interestingly, Cuvier made one mistake that is often repeated today, when in 1822 he was sent a fossilised tooth by a Sussex country doctor and fossil collector named Gideon Mantell (1790–1852). The tooth he sent to Cuvier was actually the tooth of a gigantic extinct reptile which Mantell called the iguanodon, after the large South American tropical lizards called iguanas. It would have been the first dinosaur ever identified, but unfortunately Cuvier wrongly identified some of the teeth as those of a hippopotamus. Faced with that dismissal, Mantell delayed publication. So the credit for the first description and naming of what would later become known as a dinosaur went to an Oxford don, the Reverend William Buckland (1784–1856),* who in 1824 described the bones of a 'megalosaurus' found in a quarry in Oxfordshire.

Cuvier, in line with the geological catastrophists, thought that in the past the world had been overcome by catastrophes or 'revolutions' which had swept away huge numbers of species. He did not say what these catastrophes were, but not surprisingly other geologists, like the Reverend Buckland, suggested the biblical flood as the most recent one. Interestingly, Buckland later retreated from this position in his influential contribution to the *Bridgewater Treatises, Geology and Mineralogy Considered with Reference to Natural Theology.* The biblical flood would not necessarily leave any trace in the geological record, he argued, and might not be as important in geological processes as others had assumed. Although Buckland was a heartfelt Anglican he thought that a vast geological period – perhaps even millions and millions of years – had existed before the creation of the first man, Adam.

The work of Cuvier, Mantell, Buckland and swarms of other now-eager fossil hunters was showing that there had once existed creatures on earth that had vanished long ago, and the total was rising all the time. In addition, the recognition of first mammoths and then dinosaurs showed that it wasn't just small, insignificant

---

* Buckland was certainly one of the more eccentric naturalists of the early nineteenth century, conducting his fossil-hunting expeditions in a large hat and flowing academic gown. He was famous, or rather notorious, for keeping a menagerie of animals, some large and dangerous, which he allowed to wander freely around his Oxford garden. He also decided to eat his way through the entire animal kingdom, serving up to his unfortunate guests anything from boiled sea slugs to roast panther.

creatures – maybe rough prototypes – that had vanished but huge creatures, more than a match in complexity and majesty for any alive in modern times.

As highlighted in the previous chapter, while all this fossil-hunting was not necessarily seen to support the more mechanistic transformist views, it would be a grave mistake to infer that the only other option was a literalist interpretation of biblical creation. In the early nineteenth century a number of ideas were circulating about whether new species could be formed, or how a species might become extinct. It was seen as perfectly plausible by some that there were divinely instigated natural 'laws' of creation in operation. Some believed that man was the end result of a progressive development of species. Others argued that the environment drove the steps towards humanity, with sudden changes or catastrophic events in the earth's history allowing for the unfolding of a divine plan which led to the creation of the perfect habitat for humans to inhabit.

Onwards and upwards

The third of the *Muséum National d'Histoire Naturelle* trio is the one whose name has most often been cited as a precursor to Darwin, and who has been most often caricatured by purist Darwinists for the supposed errors in his thinking – that is, Lamarck. Like Darwin's, Lamarck's legacy will also be celebrated during 2009, as this will be the 200th anniversary of the publication of his most famous text *Philosophie Zoologique*. These celebrations will probably be more significant in Lamarck's native France than in the UK, but it would be a shame for his role in the history of biology to be completely overshadowed by that of Darwin; he was after all the first to suggest developing a science of *biologie* at the beginning of the nineteenth century, and he was one of the first to present a complete and coherent transformist theory which included a mechanism for species change.

Lamarck thought that all life could be arranged in a scale from the simplest to the most complex, the roughest to the most perfect. He thought that organisms naturally improved and became more elaborate with each generation. In order to achieve this goal of perfect complexity, Lamarck insisted, nature merely needs time and favourable conditions – and time, he argued, it has in abundance. As organisms became gradually more complex, he argued, so new single-celled organisms would be continuously generated to fill the gap so that these organisms always existed at every stage of development – thereby allowing new series of species to develop. Lamarck was staunchly materialistic, following the French ethos of the time, and thought that evolution was pushed forward entirely by the mechanisms of life itself, not by any supernatural intervention. But he did think that there was a progression, a drive towards improvement, inherent in evolution. In stark contrast to his contemporary Cuvier, Lamarck did not think that species could become extinct. According to him species changed over time into new species; they did not die out.

Use it or lose it?

Of course, Lamarck's name is now inextricably linked with one of the mechanisms he suggested for evolution, now known as the inheritance of acquired characteristics or Lamarckian inheritance. The specific mechanism he is most often linked to is 'use inheritance' or 'use and disuse'. The idea has been lampooned as 'use it or lose it' and caricatured by neo-Darwinists for what they see as its apparent absurdity. I mentioned Weismann's experiments with de-tailed mice in the introduction to this book, but other caricatures refer simplistically to the absurdity of a blacksmith's son inheriting his father's strong arm or a giraffe getting its long neck by stretching to reach for leaves in higher branches over successive generations. But Lamarck's idea was actually not quite as banal or as unsubtle as this. This is worth mentioning, because it is a stick used to bash those seen as 'neo-Lamarckian' or anti-Darwinian today, and helps keep 'neo-Lamarckian' as a slur or term of ridicule among certain elements of the scientific world.

Neo-Lamarckians and neo-Darwinians alike tend to make two assumptions about Lamarck's work. First, that he believed in the direct impact of the environment on organisms which would cause them to change over time; second, that he thought organisms could adapt to the environment through their own willpower or volition. Both are misconceptions. Furthermore, it is clear from the work of Lamarck himself that they are.

Lamarck's thinking was much more sophisticated than he is often given credit for. He formulated theories on adaptation, the role of the environment and the length of time involved in transformist and geological processes, and he even touched on competitive exclusion. His theories outlined three key concepts:

1. Species vary under changing external influences
2. There is a fundamental unity underlying the diversity of species
3. Species are subject to progressive development.

Lamarck also set out two laws concerning the use and disuse of organs, and inheritance, that he speculated would explain one species transforming into another. His first law stated that 'more frequent and continuous use of any organ gradually strengthens, develops and enlarges that organ, and gives it a power proportional to the length of time it has been so used', and that conversely 'the permanent disuse of any organ imperceptibly weakens and deteriorates it, and progressively diminishes its functional capacity, until it finally disappears'. His second law was that 'the acquisitions or losses wrought by nature on individuals, through the influence of the environment in which their race has long been placed and hence through the influence of the predominant use or permanent disuse of any organ', are passed through the process of reproduction to the next generation. This is often distilled down to the caricatured descriptions of use inheritance as outlined above.

This may seem strange but there are plenty of people who still think about evolution in this way – especially, for example, when discussing the idea that our appendix or little toes could 'evolve out' because we do not use them any more (obviously this has not occurred). It is an instinctively compelling idea, and the concept of 'use and disuse' was not necessarily a new one even for Lamarck. This idea, along with that of the general transmission of acquired characteristics, was already being discussed by Lamarck's contemporaries, but he developed these into an evolutionary mechanism that adapted species to changes in their environment, similarly to the way in which Darwin later used natural selection as an adaptive mechanism.

Today it is not use and disuse, but the inheritance of acquired characteristics due to the influence of the environment, that some contemporary evolutionary biologists are now reclaiming from Lamarckism, as we shall see later. There are even some who have embraced the term 'neo-Lamarckian' rather than regarding it as the insult based on the caricature of Lamarckism that some neo-Darwinists insist upon.

Lamarck and Darwin

Lamarck's ideas had much more in common with those of Darwin than is often appreciated. Both talked about adaptive changes over a long period of time, suggesting that the earth was very old. It can be seen that Lamarck's theories did contain some strong similarities to those of Darwin: the effect of differing environments, gradual progressive change over a long period of time, and use and disuse. However there are a few crucially significant differences. For a start, in Lamarck's work a *central* role was given to the inheritance of acquired characteristics through use and disuse, as the *principal* mechanism for evolutionary change. He did not develop any theories based on the process of natural selection.

There is one other great respect in which Darwin and Lamarck were poles apart. Lamarck saw evolution as a progressive, continuous process of improvement towards greater perfection. Darwin saw it as essentially directionless. Moreover, Lamarck did not believe that creatures ever became extinct, while Darwin knew that they did.

Perhaps treated a little remotely by his much younger colleagues at the *Muséum*, in particular Cuvier, Lamarck liked to see himself as a lone voice crying in the wilderness. If Darwin was not working in an intellectual vacuum, neither was Lamarck; there were a number of transformist theorists active in scientific debates in France at the turn of the eighteenth century. However, as the science historian Pietro Corsi points out, Lamarck 'never missed an opportunity to build the myth of the isolated thinker by repeatedly denouncing the backwardness and conceptual cowardice of his contemporaries'. He goes on to suggest that 'this myth has been widely echoed in the historiographical tradition that persists to present. A methodical examination of the scientific literature of the period reveals that Lamarck, on the contrary, was fully aware of the transformist hypotheses discussed by many of his colleagues.'

Lamarck is often depicted in some historical representations as having died forgotten in ignoble obscurity. Far from it – his work was much more widely discussed than has previously been recognised. Lamarck may have been rejected by some of his scientific contemporaries – most notably Cuvier – but his legacy really lived on in the early nineteenth century as he was championed by radical thinkers after his death. While elements of his theories may not have stood the test of time, he certainly had a profound influence on the debates around the nature and origin of species leading up to Darwin.

Out of the woodwork

Back across the channel in England transformist ideas were perhaps less well developed than in France, but the mania for collection certainly got people thinking. After the publication of *On the Origin of Species* there were a number of retrospective claims as to who had come up with theory of evolution by means of natural selection first. One of the more interesting side avenues when it comes to Darwinian precursor-chasing was the Scotsman Patrick Matthew (1790–1874). Matthew's story first came to light when he wrote to the *Gardeners' Chronicle* in 1860 claiming that he had scooped Darwin by decades, coming up with a theory of natural selection in 1831, before Darwin even set foot on the *Beagle*. Darwin immediately wrote to Lyell:

> In last Saturday *Gardeners' Chronicle*, a Mr Patrick Matthew publishes a long extract from his work on *Naval Timber & Arboriculture* published in 1831, in which he briefly but completely anticipates the theory of Natural Selection.

I have ordered the book, as some few passages are rather obscure but it is, certainly, I think, a complete but not developed anticipation! ... Anyhow one may be excused in not having discovered the fact in a work on Naval Timber.

Darwin is right to say that Matthew in some ways anticipated his work, but Matthew himself acknowledged that Darwin had approached the subject in a much more rigorous manner and employed a very different philosophical approach to the theory of evolution by means of natural selection. In a letter to the *Gardeners' Chronicle* of 12 May 1860, Matthews wrote: 'To me the conception of this law of Nature came intuitively as a self-evident fact.' He continued to say that

> Mr. Darwin here seems to have more merit in the discovery than I have had – to me it did not appear a discovery. He seems to have worked it out by inductive reason, slowly and with due caution to have made his way synthetically from fact to fact onwards.

Most experts entirely believe Darwin's version of events: that he had entirely missed Matthew's suggestions, stuck as they were in the back of an obscure book on timber. Moreover, they constitute little more than an idea, with no supporting evidence. Nevertheless, Matthew's ideas are interesting because they do clearly suggest how new species may emerge through a kind of 'natural selection':

> As nature, in all her modifications of life, has a power of increase far beyond what is needed to supply the place of what falls by Time's decay, those individuals who possess

not the requisite strength, swiftness, hardihood, or cunning, fall prematurely without reproducing – either a prey to their natural devourers, or sinking under disease, generally induced by want of nourishment, their place being occupied by the more perfect of their own kind, who are pressing on the means of subsistence.

The concept of natural selection was not a new one – and Matthew was not the only person other than Darwin to suggest it – but the way in which Darwin used it was new. Ultimately, Darwin's work clearly had a more substantive impact than Matthew's. This is due in part to Darwin's efforts to provide exhaustive quantities of evidence to support his theories. But there are some key differences between their work. Among other things, Matthew did not have the same concept of a species as a population of individuals as Darwin, so for him the mechanism of evolutionary change did not depend on individual variations. Also, Matthew upheld a catastrophist version of geological thought. Thus he proposed a series of mass extinctions caused by geological catastrophes which then resulted in rapid diversification, with relatively little change in between the catastrophes. Darwin and Wallace would go on to propose a more gradual process of species change which was influenced by Lyell's uniformitarian geology.

Darwin graciously acknowledged Matthew's priority in later editions of *On the Origin of Species*, saying: 'The differences of Mr Matthew's view from mine are not of much importance … He clearly saw, however, the full force of the principle of natural selection.' Clearly delighted with this admission, Matthew had printed on his calling cards 'Discoverer of the Principle of Natural Selection', but nowadays he is virtually forgotten. My purpose here is not to argue that Matthew deserves more credit than he gets. Rather, it is to contend that numerous people across Europe and the world were chipping away at the conundrum of how new species appeared, and what the appearance of extinct types of organism in the fossil record meant. The appearance of connectedness between the species had become increasingly apparent since Linnaeus and others had created their classifications of species, and more and more naturalists were coming up with ideas about just how they were connected. At the same time, geologists were beginning to open up the earth's ancient past and give it a history, which was unimaginably long and full of variations. We clearly can no longer caricature the shift in thinking after the publication of *On the Origin of Species* as a leap from universal acceptance of 'creationism' to a battle to get evolutionary thought accepted. Indeed it is dangerous and counterproductive to do so, for it implies a level of animosity between 'religious thinkers' and 'scientific evolutionary thinkers' which is both a misrepresentation and misleading.

Even by the time Darwin was a student, some eminent thinkers accepted that life had changed through the ages, and that the world had once been thronged with remarkable creatures like the dinosaurs which were no longer alive. In other words, the picture was being gradually assembled. All it needed was the commitment and unrelenting work of Darwin to pull it into focus, and if it hadn't been Darwin, it almost certainly would have been someone else, sooner or later – and, as we shall see, most likely sooner.

Victorian pop sensation

Interestingly, Darwin was not even the first to 'go public' with evolution in Britain. That honour belongs to Robert Chambers, a journalist, traveller and author of the still-popular *Chambers' Encyclopedias*. When it was published anonymously in 1844,

Chambers' book *The Vestiges of the Natural History of Creation* created a sensation. Written in a popular style, it became an instant bestseller, running into four editions in the first year and a further eight in later years. All kinds of rumours flew about who the author might be.

As Jim Secord tells in his engaging book *Victorian Sensation*, *Vestiges* and the subject of evolution or 'development' was soon the talk of the town, chattered about over tea-tables across London. In Disraeli's *Tancred* (1847), Lady Constance comments:

> You know, all is development. The principle is perpetually going on. First, there was nothing, then there was something: then – I forget the next – I think there were shells, then fishes: then came – let me see – did we come next? Never mind that; we came at last. And at the next change there will be something very superior to us – something with wings. Ah! That's it: we were fishes, and I believe we shall be crows.*

This spoof, though comically absurd, almost sums up the scenario presented in *Vestiges*. To present his story, Chambers pulled together the latest advances in geology, natural history and embryology in order to support his vision of progressive transmutation. His tale told how everything in existence has developed from earlier, inferior forms – the solar system, earth, rocks, plants

---

* Interestingly, when Darwin published his theories well over a decade later Disraeli, then Chancellor of the Exchequer, was one of the first to publicly express horror at the association of humans and apes, exclaiming famously: 'Is man an ape or an angel? My Lord, I am on the side of the angels! I repudiate with indignation and abhorrence these new fangled theories!' Could that have been a politician speaking?

and corals, fish, reptiles, birds, mammals and ultimately man. He even speculated that man was not the end of the road, just as Lady Constance said, and that we eventually evolve into something far superior.

Chambers' version of transmutation was a wholly linear affair and made little allowances for the branching evolution or 'tree of life' which accounted for diversity in the classic Darwinian theory. According to Chambers each individual species had developed separately – there were no common ancestors. For example, two species of mammal may have evolved through similar forms in parallel, but their original ancestors would have been completely separate and never related to one another.

After 1859 *Vestiges* outsold *On the Origin of Species* right up until almost the end of the nineteenth century, so in essence it was a Victorian popular science book, but when it was first published the reaction from men of science had been decidedly cooler than its public reception. Put next to the fossil evidence, Chambers' theory looked absurd. Even Darwin wrote, in a letter to his friend Joseph Hooker dated 7 January 1845, that Chambers' 'geology strikes me as bad, & his zoology far worse'. The eminent geologist Adam Sedgwick commented severely in his review of *Vestiges* that the book 'comes before [its readers] with a bright, polished, and many-coloured surface, and the serpent coils a false philosophy, and asks them to stretch out their hands and pluck the forbidden fruit'. Interestingly, it was some of those who would later go on to champion Darwin's evolutionary theories, such as T.H. Huxley, who were most contemptuous of *Vestiges*, rather than clergymen.

It was shortly after the publication of *Vestiges* that Darwin first wrote down the outline of his theory, but it may have been the stinging reaction to the earlier book that persuaded Darwin that it would be folly to go public with his own ideas before he had built up a huge weight of evidence. It's generally assumed that he was wary of the opprobrium of the clergy, but it may in part be that one of his fears, as someone who, although a respectable authority, perhaps felt slightly peripheral to the academic establishment, was of being cut to shreds by his fellow naturalists.

Scientifically, Darwin's ideas owe little to Chambers. There was plenty of real current and up-to-date science about the history of the earth and the development of life and society available for him to draw upon. But what *Vestiges* did do was to bring the idea of evolution into the public domain, and pave the way for the relatively easy public acceptance of Darwin's much more complete and scientifically coherent theory fifteen years later. As Darwin said in the 'Historical Sketch' he attached to the third edition of *On the Origin of Species*, 'In my opinion it [*Vestiges*] has done excellent service in calling in this country attention to the subject, in removing prejudice, and in thus preparing the ground for the reception of analogous views'.

Scooped?

It seems almost inconceivable – in the light of almost a century of discovery of extinct species including mammoths and dinosaurs, more than half a century of development of transformist ideas by leading European biologists and more than a decade after the publication of a popular bestseller on evolution – that Darwin could say, as he did, that he had not come across a single naturalist who did not believe in the permanence of species. Yet Darwin, in all his writings, comes across as one of the humblest, most honest men in all the history of science, so we have to believe that he had not discussed his ideas with any naturalist who believed species to be anything but fixed. And in the same way, we have to believe the famous story of his shock when he discovered that, after decades of diligent work preparing his theory, he was about to be scooped at the last minute by Alfred Russel Wallace.

The story goes that young Wallace, a professional naturalist working in Malaysia, had a flash of inspiration after reading Malthus and in 1858 sent Darwin a brief essay containing a fairly complete summary of a new theory of evolution by natural selection. It was almost identical to the theory Darwin had been developing quietly for two decades. Completely shocked, Darwin wrote to his friend Lyell for advice, saying: 'I shall, of course, write [to Wallace] and offer to send [his essay] to any journal. So all my originality, whatever it may amount to, will be smashed. Though my Book, if it will ever have any value, will not be deteriorated; as all the labour consists in the application of the theory.' He was adamant that Wallace should be properly credited, writing later: 'I would far rather burn my whole book than that he or any other man should think I behaved in a paltry spirit.' Fate intervened when Darwin was thrown into distracted grief by the death of one of his children, and he turned the matter over to Lyell and the naturalist Joseph Hooker to decide. Lyell and Hooker arranged a neat solution in which Wallace's essay and a few extracts of Darwin's work, with date references clearly establishing Darwin's priority, were presented together under the heading 'On the Tendency of Species to form Varieties; and on the Perpetuation of Varieties and Species by Natural Means of Selection' at a meeting of the Linnaean Society on 1 July 1858. Interestingly, neither man's work attracted much attention at the time.

Wallacism

Ever since, there have been those who have tried to claim that Wallace did not receive the recognition he was due, or that Darwin in some way suppressed the younger and notably less well-off man. But Wallace himself selflessly denied that this was the case,* and most histories of Darwin accord Wallace his proper place. Indeed, Wallace became one of the most ardent champions of both Darwin and the theory of natural selection. But this common emphasis on who got there first and who really scooped whom is part of the same old pattern of hagiography, and disguises what is actually a more interesting story. First of all, there are the scientific differences between Darwin and Wallace which came to have a very profound effect on the development of what came to be called 'Darwinism', as we shall see in the next chapter. Second, the relationship between Darwin and Wallace before 1858 was actually more nuanced than is usually supposed.

Like so many men of the age, Wallace became fascinated by the latest developments in natural history in the early 1840s. Of course he read Malthus, Lyell and Darwin's *Journal* of the *Beagle* voyage – and he also read the anonymous *Vestiges*. Transmutation was the hot topic of the day for naturalists. In 1848 Wallace embarked on his first trip to the tropics with his young friend Henry Bates. Wallace spent four years in the damp heat of the

---

* For instance, on the occasion of the 50th anniversary celebrations of the publication of *On the Origin of Species* in 1909 Wallace wrote: 'Since the death of Darwin in 1882, I have found myself in the somewhat unusual position of receiving credit and praise from popular writers under a complete misapprehension of what my share in Darwin's work really amounted to. It has been stated (not unfrequently) in the daily and weekly press, that Darwin and myself discovered "natural selection" simultaneously, while a more daring few have declared that I was *the first* to discover it, and that I gave way to Darwin! ... But, what is often forgotten by the press and the public, is that the idea occurred to Darwin in October 1838, nearly twenty years earlier than to myself (in February 1858); and that during the whole of that twenty years he had been laboriously collecting evidence from the vast mass of literature of Biology, of Horticulture, and of Agriculture; as well as himself carrying out ingenious experiments and original observations.'

Amazon, collecting insects, observing monkeys and noting the differences between palm trees. Unfortunately on his return journey to Britain the boat he was travelling on caught fire and sank – destroying his possessions, including some of the specimens he had collected on his travels. When he returned to London in 1852, Wallace used his experiences to write two books and make connections in the scientific world.

The evolution of a theory

In 1854 Wallace set off to Malaysia to study nature and collect specimens. Altogether he collected 125,000 samples during his eight years in Malaysia, including 80,000 beetles and 1,000 animal species previously unknown to science. Darwin was to become one of the clients who helped fund his trip by paying for specimens of poultry and fowl. Darwin and Wallace first started to correspond with one another in 1855. All the time, Wallace was speculating about how species developed. He was conscious of how new species arose from old, writing in a paper in the *Annals and Magazine of Natural History* in 1855, 'Every species has come into existence coincident both in space and time with a pre-existing closely allied species'.

Then, in December 1857, Wallace wrote to Darwin with an outline of a theory of evolution supported by geographical evidence from his own and other naturalists' work. The only thing missing was the mechanism for evolution, natural selection. Darwin wrote back: 'Though agreeing with you on your conclusions ... I believe I go much further than you; but it is too long a subject to enter on my speculations.' Wallace must have wondered just what Darwin meant. Then a few months later, in early 1858, he was lying ill from malaria in the Moluccas in modern Indonesia when he remembered Malthus's famous book, which he had read as a young man. In what he referred to as 'a sudden flash of insight', the idea of natural selection dawned on him. Deeply excited, he quickly wrote a paper outlining his ideas and sent it off to Darwin.

Looking at the story in detail like this, it becomes clear that focusing on the idea that Darwin was a lone iconoclast, or that he was 'precursed' by some unrecognised genius such as Wallace, is actually to miss the point entirely. Many, many naturalists were thinking about how species arose at the time, and many made a contribution – not just Wallace but, of course, all the many other naturalists who corresponded with Darwin* (and some who didn't), and the countless plant and animal breeders around the world who were conducting experiments on the development of new characteristics all the time – maybe unwittingly, maybe knowing more than we imagine.

And of course, it is not an accident that both Darwin and Wallace both got the idea for their theories of natural selection from Malthus's ideas on population limits. Many other scientists were toying with similar thoughts in relation to society, which was changing so suddenly and dramatically before their eyes as the legacy of the Industrial Revolution brought astonishing technological and cultural changes. 'Progress' was the dominating theme of the day, and it is hardly surprising that everyone from economists to psychologists was thinking about how economies move, how ideas develop, how personalities evolve and much more. Four years before the publication of Darwin's *Origin*, for instance, Herbert Spencer, the man who later coined the phrase 'survival of the fittest' and who was one of the leading thinkers of his day,

---

* Darwin corresponded with nearly 2,000 different people in his lifetime. The Darwin Correspondence Project is completing the exhaustive task of compiling all of this correspondence: http://www.darwinproject.ac.uk/

wrote a book in which he outlined a theory about how the human mind had evolved in accordance with natural laws.

None of this diminishes Darwin's achievement in any way. His work was peerless, honest and insightful – and as original as that of any scientist can ever be. But he was not alone, and the story not just of how his theories came to be but of how they were received and transmitted over the next century bears this out in a surprising way, as the next chapter will reveal.

CHAPTER 3

Darwin Mutated

We are still in the dark as to all the causes of variation. We are not yet at the bottom of inheritance. We are laboring in this and the other directions, and yet the great hypothesis holds its own and is triumphant. What Goethe calls 'Thaetige Skepsis', or active doubt, has benefited the Darwinian theory, for if doubt be honest and free from prejudice, then in time the truth is sure to come.

New York Times, 15 January 1894

This comment in the *New York Times* was written in a brief review of a collection of essays by Darwin's friend and champion Professor Thomas Henry Huxley, entitled *Darwiniana*. It is, remarkably, as true today as it was when it was written, well over a century ago. Darwinism *does* still hold its own and *is* triumphant. Yet we are *still* finding out new things about some of the causes of variation and perhaps *still* do not fully understand all forms of inheritance. I am certain too that we will more closely approach the answers by entertaining honest doubt rather than sticking with entrenched positions.

It may just be coincidence, but that phrase of Goethe's, '*Thaetige Skepsis*', was also quoted by T.H. Huxley in the very first review of *On the Origin of Species* in *The Times* in 1859. In his usual fashion, Huxley hit the nail firmly on the head, championing Darwin ardently but at the same time seeing the potential flaws in his new theory. '[*Thaetige skepsis*] is doubt', Huxley wrote, 'which so loves truth that it neither dares rest in doubting, nor extinguish itself by unjustified belief; and we commend this state of mind to students of species, with respect to Mr. Darwin's or any other hypothesis, as to their origin. The combined investigations of another 20 years may, perhaps, enable naturalists to say whether the modifying causes and the selective power, which Mr. Darwin has satisfactorily shown to exist in nature, are competent to produce all the effects he ascribes to them, or whether, on the other hand, he has been led to over-estimate the value of his principle of natural selection, as greatly as Lamarck overestimated his *vera causa* of modification by exercise.' Would that some neo-Darwinists today were as open-minded and even-handed as Huxley was then!

The march of history

There is always a danger for historians of writing what Herbert Butterfield, in the 1930s, labelled 'Whig' history, after the great Victorian Whig historian Thomas Babington Macaulay who so exemplified it.* Whig history is history written from a contemporary point of view – that is, history that implies a continuous progress to the present, and identifies the 'heroes' of the past who are now revealed to have pointed to the future, and the 'villains' who stood in their way. It is an often captivating narrative that turns history into an imaginative story written in the present.† Science history has been particularly prone to this – and the story of Darwinism is a spectacular example, with Lamarck and Paley

---

* Butterfield's *bête noire* was Lord Acton, but there is no doubt that it was Macaulay who was the epitome of the Whig tendency, introducing his history of England with the observation that 'the history of our country during the last hundred and sixty years is eminently the history of physical, of moral, and of intellectual improvement'.

† This is why many historians in the 1960s began to eschew sweeping narrative histories and general analyses to concentrate on more esoteric and research-heavy topics.

labelled villains and Darwin hailed as a hero (only after his minor 'errors' have been tidied up by the better-informed scientists of the present day, of course!).

Some scientists have tended to treat accusations of whiggishness with scorn. Of course scientists know better today, they might argue. The evidence of progress in science is all around us, so why pretend otherwise? Why spoil a great story with a pedantic search for a historical perspective that's probably an illusion anyway?

In the introduction to her history of early Darwinism, *The Ant and the Peacock*, Helena Cronin (herself a neo-Darwinist)* quotes the neo-Darwinist John Maynard Smith writing about the scientist and historian Ernst Mayr's efforts to avoid whiggishness: 'I cannot imagine how a man who has striven all his life to understand nature, and who has fought to persuade others of the correctness of his understanding, could write any other kind of history [than Whig] ... Unfashionable as it may be to say so, we really do have a better grasp of biology than any generation before us.' Of course in assuming this perspective one has to allow that scientists of the future will be better informed than those of today. What then might these future scientists make of those who employ this kind of whiggish historical writing to support their own perspective today? Their approach may be as scornful as the

---

* Dr Cronin's book is incisive, yet she makes her opinions all too clear in summing up all those who objected to Darwinism with this amusing quote from the American geneticist H.J. Muller: 'It seemed to us as if he somehow couldn't understand natural selection; he had a mental block which was so common in those days.' This is quoted in a spirit of levity, but it is typical of the tendency of all too many self-professed neo-Darwinists to dismiss those in the past who happened not to agree with their modern take on evolution as just too damned stupid to get it. And if neo-Darwinists are so inclined to be wittily dismissive about other points of view in the past, doesn't it at least colour their scientific objectivity in the present?

way in which some present-day thinkers might view the work of Samuel Butler. Butler was perhaps one of the most conspicuous practitioners of 'precursoritis' in the history of evolutionary ideas. Butler felt that Darwin had not given fair or full recognition to Buffon, Lamarck or his own grandfather, Dr Erasmus Darwin, as the true fathers of evolutionary thought. He used this stance to support his own neo-Lamarckian perspective. Butler was an erstwhile acquaintance of Darwin's who had taken offence at what he felt was a slight by Darwin and his circle, and this is evident in his writing.* It is no exaggeration to describe some of Butler's work as ranting, which of course makes it great fun to read.

And, of course, Maynard Smith is partly right.† Arguably, the potted story of the background to Darwinism in this book has

---

* However, it is important not to underestimate Samuel Butler's influence when it comes to the public reception and understanding of Darwinism. He was also a well-known novelist and it is possible that because of this, his views on Darwinism were more influential they would otherwise have been. For example, he was very influential on the way in which later writers, in particular George Bernard Shaw, interpreted evolution.

† A scientist may choose to focus on a figure due to his or her own research interests and may inadvertently end up playing favourites. I would challenge any historian to deny (in private!) that they don't have the odd favourite, though we tend to be drawn to them more because of their wider roles in scientific communities, their cultural context or their personalities than because of the relevance of their respective theories today. I, for example, am fascinated by George John Romanes – sometimes called Darwin's chief disciple. This is due to my historical research interests; he was involved in the promotion of Darwin's work after his death and this is of value in trying to understand trends in scientific thought from a historical or philosophical perspective. Beyond this, if I am completely honest, one of the reasons I have maintained an interest in Romanes is not because – like Dawkins with Darwin – he supports my scientific views or my scientific research (of which I do none). Rather it is because I am fascinated by a character who believed, as he evidently did, that it was acceptable to take a number of cats from houses around Wimbledon Common

been highly whiggish! Story-telling, of necessity, always is. But that does not justify giving full rein to one's own opinions as Butler did, and as Dawkins and his some of his supporters, I feel, often do – rather than attempting to look at history objectively with an eye as unclouded by a modern scientific perspective as possible. Not even to try to do so not only risks painting a false picture of the past, but prevents us actually learning from it and gaining insights that may help avoid pitfalls in the present. And history becomes nothing more than a tool to bolster modern prejudices. That is why it will be useful to spend a little more time exploring how the Darwinian theory of natural selection was received after *On the Origin of Species* was first published in 1859 – and how it has since been filtered by a neo-Darwinist slant.

The coming of the ape man

The apparent furore that greeted Darwin's *On the Origin of Species* in 1859 is the stuff of legend, and of course the famous bout between 'Darwin's Bulldog' Thomas Huxley in the scientific corner, and Bishop 'Soapy Sam' Wilberforce in the religious corner, in June 1860 is one of the pivotal moments in the Darwinist canon. With a brilliant bit of repartee, so the story goes, the nimble Bulldog landed the lumbering carthorse a knockout blow and won the day for Darwin and science, leaving religion for dead. But it is worth further filling out the details of the story.

When Wallace and Darwin's joint papers were first read out in 1858, the lack of interest was almost deafening. For the naturalists

---

in order to set them loose several miles away to see if they had enough of a sense of direction to get home. Of course, none of them came home – Romanes seems to have quite bemused passers-by with his little experiment – although no comment is made in his letters to Darwin about how he made up for losing everyone's cats.

who had gathered to hear them, there was nothing particularly shocking or revolutionary about them – just a couple more theories on species to add to the growing number. Even the publication of *On the Origin of Species* in 1859 provoked no really strong reaction. It was only with the publication of the second edition in spring 1860 that it began to make headlines outside the scientific world.

In particular, the heat focused on the idea that humans are related to apes – even though Darwin had carefully avoided any mention of humans in his book, saying only that 'In the distant future ... Light will be thrown on the origin of man and his history.' The outraged self-professed moral authorities were adamant that they were not fooled for a moment by this omission. Whatever Darwin might or might not have said about it, they were quite certain that he really meant to link men and apes! They knew all about evolutionary theories that did so, and had for some time. And so, remarkably, even though neither Darwin nor Wallace had said *anything at all* about humans, the idea that man was descended from the apes became the dominant public image of Darwin's theories, and he was soon being portrayed as an ape in cartoons.

The usual version of the reception of Darwinism paints the publication of *On the Origin of Species* as a sudden cataclysm which shocked the staid Victorian world and sent religion into a frenzy. But it was not quite like that. Some were indeed shocked and thrown into a spiritual crisis of sorts, but they were not necessarily the majority. The younger generation at least was already too familiar with evolutionary ideas to be much perturbed – including even the clergy. Huxley later complained that 'old ladies of both sexes consider [*The Origin*] a decidedly dangerous book, and even savants, who have no better mud to throw, quote antiquated writers to show that its author is no better than an ape himself'. But it is significant that it was really only 'old ladies' (of both sexes) and a few 'savants' who were really shocked. A number of periodicals even lampooned Darwinism, the classic example being the cartoons depicting gorillas, or later portraying Darwin himself as an ape man. These, while highlighting the hot topic of the day, turned Darwin's theories into a matter of humour rather than spiritual turmoil.*

The big match

It was in something of that spirit that the big match between Huxley and Wilberforce occurred at the week-long Oxford meeting of the British Association in June 1860. The debate is often portrayed as science versus religion, but this in itself is something of a caricature. Wilberforce was by no means there as a representative of the religious view. He actually had a keen interest in science and served as a vice-president of the British Association. So Huxley was not the only voice of science in the spat. Indeed others, like Darwin's friend Joseph Hooker, may have actually played a more important part than is generally recognised. In fact, the debate over Darwin only began to gather heat that week when at one meeting Huxley stood up and directly contradicted Richard Owen, the first head of the Natural History Museum and the world's leading animal anatomist, on a small point of anatomy.

Owen had said that ape brains resembled those of the lowest monkey much more than they did human brains because, unlike human brains, they had no hippocampus. Huxley felt sure that this was wrong and said so. The audience began to sense an exciting clash of titans over the man–ape connection. In the end, over

---

* These spoofs appeared from as early as May 1860, when *Punch* published a cartoon in which a gorilla asked 'Am I a Man and a Brother?'

a year later, Huxley proved his point by showing that ape brains, like human brains, do have a hippocampus.*

Wilberforce, said to have been primed over dinner the previous evening by Owen, who was not himself present, spoke first in the debate and, according to some accounts, at one point loftily asked if Huxley was related to apes on his grandmother's side or his grandfather's. At that, Huxley is said to have whispered to his friends triumphantly, 'The Lord hath delivered him into mine hands.' Huxley's riposte was that if the question were put to him, 'would I rather have a miserable ape for a grandfather or a man highly endowed by nature and possessed of means of influence and yet who employs those faculties for the mere purpose of introducing ridicule into a grave scientific discussion – I unhesitatingly affirm my preference for the ape!' The audience, according to some reports, roared with laughter and clapped thunderously, and Huxley wrote to a friend: 'I was the most popular man in Oxford for a full four and twenty hours afterwards!' Interestingly enough, Wilberforce himself felt that he had won the day and was afterwards always most cordial to Huxley. No full account of this meeting exists but a number of reports survive from both camps. Wilberforce's review of *On the Origin of Species*, published in July

---

* So famous did this exchange become, and such a source of amusement even to clergymen, that Charles Kingsley could allude to it in his children's fantasy *The Water Babies*: 'The professor ... held very strange theories about a good many things. He had even got up once at the British Association, and declared that apes had hippopotamus majors in their brains just as men have ... If you have a hippopotamus major in your brain, you are no ape, though you had four hands, no feet, and were more apish than the apes of all aperies. But if a hippopotamus major is ever discovered in one single ape's brain, nothing will save your great-great-great-great-great-great-great-great-great-great-great-greater-greatest-grandmother from having been an ape too.'

1860, gives us some insight into his stance on Darwin's theories, and it is interesting to note that he criticised them from both a scientific and a philosophical point of view, commenting that he had 'objected to the views with which we are dealing solely on scientific grounds'. He went on to say that he had 'no sympathy with those who object to any facts or alleged facts in nature, or to any inference logically deduced from them, because they believe them to contradict what it appears to them is taught by Revelation.'

The debate between Huxley and Wilberforce was clearly not a simplistic debate about the clash between religion and evolutionary theory. It was distinctly political, not least because Huxley was one of a new wave of 'professional' empirical men of science seeking to replace the unpaid 'gentlemen of science'. He used this occasion to attack a member of what he saw as the old, financially independent ecclesiastical elite that he believed barred the way to those who sought to create a career in science – as did those who wrote about the debate in later years. Then, as now, it is impossible to take the politics out of the ostensible rift between religion and science.

Missing links

While we often hear about those who responded with shock or outrage to Darwin's theories, there were from the outset a number of people who found no problem with accommodating Darwinism alongside their religious beliefs. A classic example of this was Sir Edward Fry, a member of the famous chocolate-producing Quaker family. Fry had never felt that Darwinism was inconsistent with the teachings of the Bible, and he happily embraced evolutionary theories. Fry commented: 'I have no fear whatever, of further investigations into nature; I have no fear of true science, though I have much fear of false science, and of false theology too.'*

Churchmen withdrew their opposition to evolution with what, by the standards of often repeated accounts, was astonishing rapidity. Darwin's *Descent of Man*, in which he did finally link men and apes, met with only a small ripple of controversy when it was published in 1871. By the 1880s, Huxley was left grumbling that there were no more theologians to tackle: 'There must be some position', he lamented, 'from which reconcilers of science and Genesis will not retreat.' And when Darwin died in 1882, the father of evolution was respectable enough for his colleagues to ensure that he was buried in Westminster Abbey, the most honourable last resting place the Anglican church could bestow.

It was not necessarily theologians who proved the most persistent critics of Darwin's theories in the long run, but his fellow naturalists. For a while, a few in the science community including Huxley remained sceptical of the gradual process of evolution by means of natural selection. Huxley did not of course reject evolution per se; rather, he proposed a version of evolution known as 'saltationism'. Its central argument was that evolution occurs by a series of leaps or jumps between species, not in the gradual way that Darwin had proposed.

It is suggested that during the 1850s Lyell and Huxley discussed the problems with Darwin's gradual view of evolution, and how it apparently failed to account for either the development of new

---

* Fry had eagerly read a copy of *On the Origin of Species* in 1861, and in 1872 he went on to publish a book entitled *Darwinism and Theology*, a reprint of letters originally published in the *Spectator* where this comment was made. Fry, who had a keen interest in botany and zoology, published three books on mosses, liverworts and slime moulds. He later went on to become the Lord Justice of Appeal and the vice-president of University College London.

species or the gaps in the fossil record – a problem they both also had with Lamarckian evolution. These were real scientific doubts from the most respected men of science at the top of their game, not the blinkered, religiously slanted objections of an older generation – despite Huxley's candid admission, when he heard the full details of Darwin's theory of natural selection: 'How stupid of me not to think of that!' Again this is a debate that has lingered into the twentieth century – Stephen Jay Gould's concept of evolution by punctuated equilibrium shares some similarities with the saltationist theories of Huxley and others. Gould's punctuated equilibrium, along with his adherence to a kind of group selectionism, were key features in his ongoing debates with Dawkins and Daniel Dennett, as we will discuss in the next chapter.

The science historian Peter Bowler, in his book *The Non-Darwinian Revolution* (1988), argued that the best-known aspects of Darwin's theories did not form the key thread of evolutionary science in the later nineteenth century or even the early twentieth century. By his analysis, 'the Darwinian revolution was not such a one-man show after all. Darwin may have hit upon the explanation of evolution favoured by modern biologists, but he obviously encountered major difficulties in presenting it to his contemporaries.' Bowler later clarified that he did not seek to discount the impact of Darwin's work – which was clearly significant – but that there was a need for it to be reassessed. While some historians and modern biologists may focus on the history of evolution through a modern neo-Darwinian perspective, this fails to take into account the important role of work which might now be classified as 'anti-Darwinian', or the impact of evolutionism in other spheres of knowledge. Here I agree with Bowler, and would further argue that our perspective on what constitutes 'anti-Darwinism' is a construct of some of the previous histories of Darwinism and their mostly inadvertent neo-Darwinian slant. It is important to put Darwinism into the context of the wider evolutionary debates. The period at the turn of the nineteenth century is sometimes referred to as the 'eclipse of Darwinism' – a phrase coined by Thomas Henry Huxley's grandson, the biologist Julian Huxley. This is because, far from Darwinism, or rather evolution by means of natural selection, being instantly or even gradually accepted, there were a number of competing evolutionary theories – some of which significantly downplayed the role of natural selection or saw it as subordinate to other progressionist mechanisms.

The selection problem

In his address to the British Association meeting in 1868, Hooker suggested that Darwinism was 'an accepted doctrine with every philosophical naturalist'. However, Hooker admitted that there were 'a considerable proportion who are not prepared to admit that it accounts for all Mr Darwin assigns to it'. Darwinian *evolution* was to become widely accepted within a few decades. Yet *natural selection*, Darwin's principal mechanism for evolution, was not so widely accepted, and this was by scientists rather than clergymen. The beauty of natural selection as an idea is that it is entirely automatic and mechanical. As Dawkins has shown so eloquently in books such as *The Selfish Gene* and *Climbing Mount Improbable*, natural selection alone is potentially capable of accounting for all the teeming variety of life on earth without any divine intervention. Interestingly, this was less of a problem for churchmen in the nineteenth century than it seems to be now – this is in part because the more worrying implications were reconciled with progressive or goal-directed versions of evolutionism. Another factor in the lesser acceptance of natural selection was that there were a number of questions that were seen as posing considerable problems. These can be broken down into six main points, which were fairly well outlined in 1871 by one of Darwin's more vocal critics, St George Mivart.*

The first of Mivart's points was that there were structures that existed that were either non-adaptive or apparently useless. The second was that there were seemingly similar structures that had evolved in apparently unrelated species. For example, the eyes of cephalopods† are fairly similar to vertebrate eyes. The third was the matter of the saltationists' challenges to the gradual process described by Darwin.

The last three are of most interest to us.

The fourth was the debate surrounding the mechanisms by which variation and inheritance could occur. The fifth was the lack of data from the fossil record. And the final point, and perhaps the most problematic for Darwinians in the late nineteenth century, was the length of time needed for Darwin's gradualist evolution by minute steps to work, which required the earth to be at the very least millions of years old.

Darwin responded to all of these criticisms at the time, but they are still claimed to have had an impact on the development of his theories, especially his theory of inheritance. The debates that are most often cited as having influenced Darwin's concep-

---

* St George Mivart was a Catholic anatomist and zoologist, who had been one of the earlier supporters of Darwin but later became gradually more dissatisfied with the theory of evolution by natural selection. This culminated in 1871 with the publication of *Genesis of Species*. *Genesis* was not just an outline of Mivart's own criticisms of Darwin's work but a comprehensive collection of all of the most incisive criticisms of Darwin's work to date; further to this, it highlighted the conflicting opinions of the Darwinians themselves.
† The class of species that includes octopuses, squids and cuttlefishes.

tualisation of heredity were those around variation and the age of the earth.

Too young for evolution

The challenge to Darwin's assumption of a really ancient earth came quite quickly, and it was from a source that could not easily be dismissed – William Thomson (later Lord Kelvin). Thomson was one of the country's most prominent physicists, with a reputation for exactitude. It was not actually Darwin whom he challenged directly, since he was a believer in evolution, but Lyell and his gradualist (or Uniformitarian) picture of the earth's history, which demanded an ancient earth. In 1866, Thomson published a paper entitled 'The Doctrine of Uniformity in Geology Briefly Refuted' in which he argued that the earth was much, much younger than Lyell and his colleagues suggested.

Like Buffon a century earlier, he based his case not on geology but on physics, assuming that the earth had begun as a hot molten ball, then calculating how long it would have taken to cool to its present temperature. He proposed that, based on the calculations available at the time for the length of time needed for the earth to cool to its present temperature, the most probable figure for its age was 100 million years, with an upper boundary of around 400 million. To make matters worse, he later reduced these figures even further. These estimates simply did not allow for evolution by natural selection as Darwin had envisaged it. Huxley leaped to Darwin's defence with a lecture attacking Thomson's assumptions, but other physicists such as Helmholtz and Newcomb backed up Thomson's lowest figure. Although Thomson did not disagree with the idea of evolution per se, he did believe that there was evidence for design in nature. He argued that his limitations on the age of the earth had effectively debarred the possibility of natural selection as a mechanism for evolution.

It was not until the early twentieth century and the discovery of radioactivity, which showed that the earth's core continues to generate heat, that Lord Kelvin's estimates were roundly discounted. Current radiometric estimates put the earth at just under 4.6 billion years old, and the emergence of life at well over 3.5 billion years ago, more than enough for Darwinian evolution. But back in the nineteenth century, there is no doubt that faith in Darwin's vision of evolution was seriously weakened by Thomson's calculations.

The case for evolution was already solid, but it was the mechanism by which evolution happened that was in doubt. If the earth was as young as Thomson asserted, evolution would have to move quickly, much more quickly than natural selection could feasibly do – especially since a young earth would have a turbulent history which would constantly interrupt the process. That is why mechanisms like Huxley's saltation, orthogenesis (the idea that organisms had some natural internal drive to evolve) and even neo-Lamarckian evolution could not be dismissed, because they all worked much faster.

Mixed blood

Within just a year of the blow about the age of the earth, Darwin faced another serious scientific criticism. This came from a close friend of William Thomson, the then professor of civil engineering at University College in London called Fleeming Jenkin (1833–85). Jenkin pointed out a real problem with Darwin's contention that beneficial adaptations would persist through the generations. Jenkin is now, perhaps unfairly, best known for having publicly put forward the idea of 'blending inheritance' as an argument against Darwin's natural selection theory. In his 1867 review of *On the Origin of Species* in the *North British Review*, Jenkin outlined various arguments against Darwinism. Among other things he suggested that there were limits to variation, disputed the possibility of common ancestry and upheld Thomson's estimations of the age of the earth.

However, the key point that has been picked up by historians and scientists today is the idea of a 'swamping effect', or blending inheritance. Jenkin argued that a single abnormal variation would be lost in successive generations, even if it were beneficial to the species. Any single advantageous individual variation or character would, when introduced to a wider and less advantaged population, be swamped by the inferior characteristics during the process of breeding. A common analogy to Jenkin's argument is the idea of mixing a spot of white paint into a pot of black paint.*

This obviously assumes that there is no transmission of favourable variations beyond one generation, whereas modern theories of inheritance (through the transmission of both recessive and dominant genes) do allow for transmission to successive generations.

In the nineteenth century, however, the problem of the 'swamping effect' was a theoretical issue that Darwin had to overcome. It

---

* According to Jenkin, for a variation to remain part of a species it would have to occur in a large enough proportion of the population to be maintained over subsequent generations. New variations occurring in a small number of members of the species would be swamped or blended out. Therefore there was a limit to the number of variations that would be maintained. Jenkin also argued that while in artificial selection the breeder could force individuals with an advantageous characteristic to breed with one another, thus maintaining that characteristic, in the wild there would be no control over the breeding habits of a species; it would be highly unlikely that the individuals who had inherited an advantageous characteristic would happen to breed with each other.

is often argued that this caused him to rely on the inheritance of acquired characteristics, and particularly the mechanism of use and disuse, more heavily in his later work.

Use and disuse went some way towards countering Jenkin's argument. To use the oft-cited example of the giraffe's neck, consider what would happen if an ancestor giraffe with a long neck were to breed with a short-necked member of the species which had managed to survive by eating lower leaves. If Jenkin were right, the inherited effect of the elongated neck would be lessened by that of his mate's shorter one – their offspring would have medium-sized necks, and over successive generations the descendants' necks would get shorter and shorter until all advantage was lost.

However, if the ancestor giraffe developed its long neck over a lifetime of using it to reach for high leaves – and if, as is likely, more than one individual attempted to do this – then several members of the species would be producing medium-necked offspring. Therefore, there would be enough of the latter for more medium-necked animals to interbreed with each other. If each successive generation of medium-necked animals were to reinforce this longer neck trait and continue to breed with each other, then over a relatively short number of generations a population of long-necked individuals would emerge. Arguably, this kind of inheritance of acquired characteristics could shorten the period of time that life would need to evolve. Of course, this version of evolution leads to some problematic inferences – namely a kind of teleological process that means that a species could theoretically control its own evolution, or that there is a definite end point or direction to evolution. This is an aspect of such mechanisms which has dogged the history of evolution and, as we will see later, provides an interesting insight into Dawkins' perspective on science and religion.

As we can see, resorting to use inheritance could arguably reduce the length of time needed for evolutionary processes and counter the problem of the swamping effect. However Darwin never really accepted the shorter version of the earth's history, and stuck to his gradual version of natural selection. Of Darwin and Wallace it was Wallace, who rejected outright any form of inheritance of acquired characteristics, who was more confident that natural selection could work on a much shorter timescale.

It is rather ironic that Jenkin sums up his paper by saying: 'What can we believe but that Darwin's theory is an ingenious and plausible speculation, to which future physiologists will look back with the kind of admiration we bestow on the atoms of Lucretius, or the crystal spheres of Eudoxus, containing like these some faint half-truths, marking at once the ignorance of the age and the ability of the philosopher.'

Without Mendel

Some histories of Darwinism propose that if only Darwin had been aware of Mendel's work on peas, which is widely seen as the starting point of modern genetics, he would have had no problem with blending. Michael Ruse wrote ruefully: 'If only Darwin had known that the way was being opened to defuse the Jenkin criticism entirely.' In fact, Darwin had his own ideas on how inheritance might occur without blending which actually predate Jenkin's criticism by decades. Although they were published only in 1868, the manuscript had been finished some time before Jenkin's critical review of *On the Origin of Species*. And although virtually ignored until recently, Darwin's ideas on heredity are beginning to be revisited in light of current developments.

Darwin saw that the inheritance mechanism had to involve particles that preserve variations from generation to generation. Particles were also needed, Darwin realised, to solve the problem of atavism, the reappearance of ancestral traits several generations later. These particles sound a little like genes. Indeed the word 'genes' comes from Darwin's name for his particulate process, 'pangenesis'. In an essay in *The Devil's Chaplain* Dawkins suggested, before entirely rejecting them, that Darwin's ideas were a tantalising precursor to Mendel. But Darwin's concept of pangenesis is very different from Mendelian inheritance, and if Darwin had ever read Mendel's work he would probably have discounted it.

Darwin realised that his ideas were entirely speculative, commenting in the 1868 first edition of *The Variation of Animals and Plants under Domestication* that

> As Whewell, the historian of the inductive sciences, remarks: 'Hypotheses may often be of service to science, when they involve a certain portion of incompleteness, and even of error.' Under this point of view I venture to advance the hypothesis of Pangenesis, which implies that the whole organisation, in the sense of every separate atom or unit, reproduces itself. Hence ovules and pollen-grains – the fertilised seed or egg, as well as buds – include and consist of a multitude of germs thrown off from each separate atom of the organism.

It is easy, from a modern perspective, to read cells and DNA into his statement that 'the whole organisation, in the sense of every separate atom or unit, reproduces itself'. After all, we know today that every body cell contains DNA, the complete instructions for making a new organism. Modern classical genetic theory, though, says that the body cells are separate from the germ cells involved in reproduction, so only the DNA in the germ cells is actually ever used to make a new organism.

The genuine particle

What Darwin proposed were countless minute particles which he called 'gemmules'. Gemmules, Darwin imagined, were released by each part of an organism at every stage of its development, and each of them could reproduce an exact copy of the part from which it was released.* He surmised that they must have a natural affinity which drew them together from all corners of the body to the sexual organs, or the areas where asexual reproduction took place. They not only played a part in the production of the next generation, but 'latent gemmules' could lie dormant and intact in order to be transmitted to future generations. So according to this theory it was not only the reproductive organs or buds of an organism that gave rise to a new organism, but every part or

---

* Darwin was at pains to point out that, while it may seem inconceivable to some that such minute gemmules could exist and be spread through the organism in such a fashion, the work of William Thomson (Lord Kelvin), with whom his son George Darwin actually worked (and who had done so much damage to Darwin's theory with his estimate of the age of the earth), supported just such an idea:

> From data arrived at by Sir W. Thomson, my son George finds that a cube of 1/10,000 of an inch of glass or water must consist of between 16 million millions, and 131 million million molecules. No doubt the molecules of which an organism is formed are larger, from being more complex, than those of an inorganic substance, and probably many molecules go to the formation of a gemmule; but when we bear in mind that a cube of 1/10,000 of an inch is much smaller than any pollen-grain, ovule, or bud, we can see what a vast number of gemmules one of these bodies might contain.

unit of the parent organism. What was beautiful about this idea was that it not only accounted for inheritance and asexual reproduction but also variation, since gemmules could lie dormant then reappear, and because new gemmules could be released in response to changing environmental conditions or through the use and disuse of parts.

To give an illustration of this idea, begonia plants can reproduce either sexually through pollination of the ovum, or asexually through leaf segment cuttings. Under pangenesis, gemmules from all parts of the plant would be collected in the leaves and in the sexual organs, so there would be different gemmules available for reproduction from stalk, roots, flowers etc. These gemmules might not be expressed in every generation. If two parent plants were crossed, say one with white flowers and one with red, a blending effect would give pink flowers; theoretically with pangenesis occasionally a white flower may appear instead. A white flower might then be present in the next ten generations and then, in generation eleven, we might see a red-flowered plant. This is because the red flower gemmules have not been lost, but are latently stored and passed down the generations.

Passing on

Now this might sound a little like modern classical genetics – but there are two significant differences. First of all, Darwin did not propose any mathematical laws as to how different gemmules may be expressed – an essential feature of genetics – although there are some key hints in his work towards mathematical ratios. Second, and more significantly, pangenesis allows for a number of kinds of inheritance of acquired characteristics, because gemmules could be developed and released *during an organism's life* in response to changing conditions. This contrasts entirely with genetic theory, in which the genes involved in reproduction *are never altered* from generation to generation; they are merely exchanged during sexual reproduction or mutated during the reproductive process. According to the usual representation of Weismann's work, which is seen as a step towards genetic theory, only the germ cells are ever involved in inheritance. In pangenesis, cells from the whole body may be, hence the name *pangenesis*.

There are two main points that we should note about Darwin's pangenesis. First, the hypothesis applies mainly to plants and lower life forms. Second, as Darwin states categorically in *Origin*, mutilations cannot be inherited, something he later reiterated in relation to pangenesis. So the oft-repeated assertion that Darwin's pangenesis was dismissed by Weismann's experiments of cutting off the tails of mice to see if they would produce tailless offspring is erroneous.

So Darwin's gemmule theory was a particulate process that as well as accounting for adaptation by natural selection also allowed for use and disuse, and for an organism's environment to play a major part in its evolution. That is why it has been written off by many neo-Darwinists as a Lamarckian dead-end – or worse, the desperate attempts of an ageing man to fend off rising criticism of his theories which were correct only in their younger, purer form.

There are two important issues to address here. The first is to make clear that Darwin began developing these ideas almost as early as he began developing the idea of natural selection. So it is misleading of neo-Darwinists to claim that the modern synthesis of genetics and natural selection is 'pure Darwinism'. Secondly, as current trends in research change, it is just possible that Darwin's ideas on pangenesis might not have been so far off the mark as the neo-Darwinists assume. As we shall see, there has been a renewed interest in Darwin's gemmules in light of recent research in epigenetics, which enables inheritance by other means than genes. Ironically, in this respect it might just turn out that Darwin was more on the right track than Dawkins is today.

Dawkins approaches evolution from an ethological or zoological perspective. These disciplinary distinctions may seem minor; after all, they all relate to the same evolutionary mechanisms. But some who work in areas like the plant sciences or microbiology have long been aware that a strict neo-Darwinian, gene-centric model of evolution might not quite fit with what is observed in nature.

'Inheritance of acquired characters', both at this point and throughout the late nineteenth century, was an umbrella term that referred to a number of mechanisms, including the effect of the environment external to the organism and use and disuse (also known as use inheritance). Use inheritance is what is usually known as 'Lamarckian inheritance'. There was much debate in the late nineteenth century about what exactly is meant by the term 'acquired characters', and during the twentieth century the term was used to mean a number of things. For clarity's sake – as the phrase is still used in contrasting ways today – when I refer to the 'inheritance of acquired characteristics'* I mean all the mechanisms to which this can refer, not merely use and disuse.

The Wallace selection

The whole idea of pangenesis shows that Darwin was far from the pure natural selectionist that neo-Darwinists often assume he was, and that he even softened his line over the years. It was

---

* In the nineteenth century people tended to refer to the 'inheritance of acquired characters'; from the beginning of the twentieth century the phrase 'inheritance of acquired characteristics' became more common.

Wallace, however, who remained the strictest selectionist – though he was not a great fan of the term 'natural selection' which he thought was open to misinterpretation, implying that there was active selection at work which could involve a divine hand. It was Wallace who urged Darwin to use Spencer's term 'survival of the fittest' instead, because it was purely a description of the process.

However, there were a few differences between Darwin and Wallace's views of natural selection. Particularly later in their careers, by the 1870s, the two men had begun to part ways on a number of issues including the origin of man as an intellectual and moral being, pangenesis and the inheritance of acquired characteristics. Wallace was convinced of the 'overwhelming importance' of natural selection, and after initially being open to the idea had turned against Darwin's pangenesis.

A topic of much debate is whether, when the two first formulated their theories, they differed on what was being selected. For Dawkins, it is the gene. For Darwin, it was the individual organism. For Wallace in 1858, it has been argued that it was varieties or sub-species.

In his 1858 paper that was presented jointly with Wallace's, Darwin used the example of foxes catching rabbits, arguing that 'those individuals with the lightest forms, longest limbs, and best eyesight, let the differences be ever so small, would be slightly favoured, and would tend to live longer, and to survive during that time of the year when food was scarcest; they would also rear more young, which would tend to inherit these slight peculiarities.' To sum up: 'The less fleet ones would be rigidly destroyed.' Wallace, too, described in his 1858 paper how pigeons with shorter wings (which were therefore able to fly less far to find food) and antelopes with shorter legs (which would be more easily caught by fast-running predators) might fall by the wayside. But Wallace argued that it was *varieties* of organisms that were selected, saying that if 'any species should produce a variety having slightly increased powers of preserving existence, that variety must inevitably in time acquire a superiority in numbers.'

Darwin's focus on the individual was one of the most important aspects of his work, as it was the first theory of transmutation to propose a feasible selectionist process of evolution based upon individuals within a population (as opposed to wholly progressionist theories, like that proposed in Chambers' *Vestiges*).

This had been highlighted as early as 1894 by the geologist and palaeontologist Henry Fairfield Osborn, in his book *From the Greeks to Darwin*. Edward Bagnall Poulton, a long-term supporter of Wallace's, later recounted what Wallace himself had said about this whole debate:

> I used the term 'varieties' because 'varieties' were alone recognized at that time, individl [*sic*] variability being ignored or thought of no importance. My varieties therefore included individual variations.

This is an interesting historical point, but it has had little impact on the way in which Darwinism is represented today. Wallace later changed his perspective and came to think that individuals were the unit of selection – to some degree he emphasised this more strongly than Darwin in the end. However, in the recent Open University annual lecture, Dawkins recounted the argument about levels of selection that has been put forward by historians of science, 'that Wallace's version of natural selection was not quite so Darwinian as Darwin himself believed'.

In a stunning piece of presentist whiggishness, Dawkins then went on to say: 'Some people have suggested that Wallace, unlike

Darwin, who clearly saw selection as choosing among individuals, was proposing what nearly all modern theorists rightly denigrate as group selection.' He later continued: 'But a careful reading of his paper rules it out. By "variety" and "race", Wallace meant what we would nowadays call genetic type, even what a modern population geneticist might mean by an allele.'

Why does all this matter – what difference does it make if Wallace and Darwin thought differently about evolution or variation? To some degree it is a point of mere historical debate, though it clearly meant a lot to Wallace, Poulton and latterly Dawkins. Moreover, this is not surprising, because when it comes to what we think of as Darwinism today – as we shall see in the next chapter – Wallace quite literally wrote the book!

The most acute differences between Wallace and Darwin, however, became more prevalent later in their careers. They parted company on a number of issues including not only the inheritance of acquired characters, but also sexual selection and the origin of the human mind through non-spiritual means. Nonetheless, Wallace insisted on the logic of natural selection. Darwin's reliance on the inheritance of acquired characteristics is often written out of history books. However, Darwin himself never abandoned his theory and felt that it would eventually be accepted and proved by future research. In his autobiography, Darwin said that:

> My *Variation of Animals and Plants under Domestication* [...] gives all my observations and an immense number of facts collected from various sources, about our domestic productions. In the second volume the causes and laws of variation, inheritance, &c., are discussed, as far as our present state of knowledge permits. Towards the end of the work I give my well-abused hypothesis of pangenesis.

An unverified hypothesis is of little or no value; but if any one should hereafter be led to make observations by which some such hypothesis could be established, I shall have done good service, as an astonishing number of isolated facts can be thus connected together and rendered intelligible.

What is interesting is that it was Wallace's rigid natural selectionism that became the orthodox face of Darwinism, and this involved rejecting some of Darwin's own ideas. Current debates in evolutionary biology that challenge the notion of evolution working solely by natural selection of genes are presented as though they are challenges to Darwinism. In fact they are challenges to 'neo-Darwinism', a slightly derogatory term coined by Darwin's friend George Romanes in the 1890s to describe the theories of Wallace and, more particularly, August Weismann, of whom more later. If some of the staunch neo-Darwinists of today were to take a leaf out of Darwin's book, they might be more open to some of the exciting new research in evolutionary biology, developments that Darwin himself might surely have understood and, given his immense curiosity and open-mindedness, would almost certainly have found fascinating.

The negative reception of Darwin's theories is usually depicted today as a reaction from a creationist worldview that was born out of old-school theological perspectives. Any nineteenth-century challenges to his ideas are now dismissed either as blind alleys that could have been avoided if Darwin had been aware of Mendel's work – the precursor of modern classical genetics – or as inadequate attempts to either reconcile or dismiss natural selection with a more teleological or directed process. While some of his critics – Mivart and Jenkin included – were arguing for a more definite role for a divine power in the evolutionary process, their challenges were not seen in specifically theological terms. Indeed, Jenkin's criticisms – while clearly unsubstantiated from a modern perspective – were based on sound science that was grounded in the best evidence of the day. We need to understand these challenges on their own terms, rather than retrospectively using them to argue that there was a significant divide between 'science' and 'religion', 'natural selection' and 'teleology'. After all, one of the people who played a large part in promoting natural selection, Wallace, himself believed that the human mind was not a product of evolutionary processes. Even the most oft-cited battle between 'science' and 'religion', the now infamous debate at the British Association in 1860, today tends to be represented in a caricatured way that fails to take into account the concerns that really fuelled the debate. Darwin was indeed criticised from a theological perspective, but the picture is much more complex than it first appears. It is impossible to remove the science from its ideological context.

Ironically, the lament that Darwin was pushed into accepting a 'Lamarckian' version of evolution towards the end of his life has led to the overlooking of one of his more interesting hypotheses. And while it would be equally whiggish to claim Darwin's pangenesis as a precursor to modern epigenetics, it certainly raises some interesting questions as to what else has been written out of the history of evolution. If the history of science focuses only on those thinkers or theories that are seen to be acceptable by modern standards, it gives us a very skewed perspective of the trends that have developed since then. It is all too easy to paint a picture that seems to show that we have marched inevitably towards a modern gene-centric model of evolution. But is this really the case? This kind of progressionist history can serve only to reinforce the perspective of researchers today. In some ways it can act to reinforce dogma and constrain debate – after all, nobody wants to be labelled as being part of an anti-Darwinian or a Lamarckian tradition.

CHAPTER 4

WHOSE DARWINISM IS IT ANYWAY?

Darwinism is the differential survival of self-replicating genes in a gene pool, usually as manifested by individual behavior, morphology, and phenotypes.

Richard Dawkins in an interview with
Frank Miele in *Skeptic* magazine, 1995

Challenges to the orthodox neo-Darwinian synthesis are sometimes presented as challenges to Darwinism. Similarly, neo-Darwinists conflate their own ideas with Darwinism. That's why Dawkins feels quite at ease saying that 'Darwinism is the differential survival of self-replicating genes in a gene pool, usually as manifested by individual behavior, morphology, and phenotypes.' But this is Dawkinism, of course, not Darwinism. Darwin had no notion of 'the differential survival of self-replicating genes in a gene pool'. It's not just that Darwin didn't know about genes; his idea of natural selection was significantly different.

First of all, there is the obvious difference that Dawkins and Darwin envisage different units of selection. Dawkins, of course, plumps for the gene, while Darwin focuses on the individual organism. But there is more to it than that. Dawkins' idea hinges on the immortality, the survival of the gene, whereas Darwin's version is based on the transmission of variation. In Dawkins, the starting point – the selfish, replicating gene – must persist, however much the organism – its 'vehicle', as Dawkins describes it – changes. In Darwin, variation is driven both directly and indirectly by the environment in which the organism lives. So Dawkins' version of evolution can work *only* with natural selection, because if genes are ever shown to vary like organisms do – if they do not persist, if they are not in essence immortal – then they can no longer be considered selfish or self-replicating. (That is why Dawkins threatened to eat his hat if signs of Lamarckian inheritance could be proved.) Darwin's version, on the other hand, while driven mainly by natural selection, does not necessarily exclude other mechanisms and does, as it happens, allow the particles of inheritance to vary.

Neo-Darwinism, then, is a hybrid that is only partly based on Darwin's ideas, and Darwin would not necessarily have agreed with it or even recognised it. It is not merely the knowledge of genes and DNA that developed during the twentieth century that makes it different; it is the entire vision of evolution. In fact, neo-Darwinism springs more directly from Wallace and August Weismann than from Darwin. Indeed, had Darwin *never existed*, we might have ended up more or less where we are today with neo-Darwinism. Wallace provided the idea of natural selection and like neo-Darwinists was a strict selectionist, while Weismann apparently introduced the idea of the isolation of the germ line that invalidates any other method of evolution *but* natural selection. It's because Wallace and Weismann's ideas were *not* strictly Darwinian that Darwin's friend George John Romanes called them *neo*-Darwinism, not because they were.

Weismann's barrier

By a strange and perhaps telling coincidence, the shift to neo-Darwinism began in Britain the year Darwin died in 1882, with the publication of the translations of Weismann's *Studies of the Theory of Descent*. This was followed in 1893 by the same author's *The Germ Plasm: A Theory of Heredity*. Weismann was a German cytologist who criticised Darwin's view of pangenesis from the perspective of the then newly developing understanding of cell structure. Weismann started to put together his own speculative theories on inheritance, based partly on his previous work on the structure of the cell nucleus. He suggested that the cell nucleus was responsible for inheritance and proposed that there was a distinct separation of the 'soma' (or body cells) of an organism and its 'germ line' (or sex cells).

This is crucial because although, according to the usual portrayal of Weismann's work, the soma cells can be modified during an organism's life, these modifications are never passed on to the sex cells; the material of the sex cells, or 'germ plasm' as he called it, is passed on intact down through the generations. So offspring come into the world just as their ancestors did, unless the germ plasm itself is altered. According to this view, the variation that results in evolution can occur only through direct changes to germ plasm – that is, either spontaneous random mutations or sexual exchanges. This permanent barrier between the soma (body) cells and the germ plasm (sex cells) – now known as the Weismann barrier – means that there can be no 'Lamarckian inheritance'. Weismann, of course, apparently proved his point with the gruesome experiment in which he cut off the tails of countless mice to show that their taillessness was not passed on to their offspring.*

Though Weismann's exact mechanisms of heredity have been superseded and disproved on a physical molecular level, he is still

---

* This is often seen as the archetypal disproof of the inheritance of acquired characters and, by proxy, pangenesis. However, as Darwin himself would have been the first to point out, this shows only that mutilations are not inherited. Where would be the adaptive advantage in a mutilation being inherited? Clearly, this does rightly discount one of the more naïve forms of the concept of inheritance of acquired characters. However, it does not discount other forms of this type of inheritance.

credited with the theoretical developments that led to the exclusion of Lamarckian principles, or use inheritance, from serious evolutionary thought.* However, the key point here is that he discounted use inheritance but not all forms of the inheritance of acquired characteristics. Given that Weismann is seen as the archetypal neo-Darwinian, a doctrine which at its heart holds that there can be no impact from the environment on any unit of heredity, it is surprising to find that – in the words of the philosopher of biology Rasmus Winther – Weismann was no Weismannian!

From 1894 onwards Weismann started to develop his theory of 'germinal selection'. This theory suggested that the germ plasm *could* acquire changes during the lifetime of an organism which are then inherited. Strange, then, that Dawkins has claimed – for instance in *The Extended Phenotype* – to be a kind of Weismannian.

Dawkins obviously uses Weismann's theories in his definition of 'Darwinian', as he refers to the 'immortal germline'. Dawkins' definition of germline is linked to its immutability. For him, the implication is that the germline, like his replicators, is in essence immortal. But this is a naïve reconstruction of Weismannism, as Weismann himself actually denied the idea of a permanently

---

* The similarity of Weismann in his earlier work to that of modern-day ultra-Darwinists is not limited merely to the exclusion of the inheritance of acquired characteristics. Although his views are tempered by a top-down perspective, Weismann sometimes comes very close to the reductionist line used by Dawkins in his selfish gene theory:

> ... from the point of view of reproduction, the germ-cells appear the most important part of the individual, for they alone maintain the species, and the body sinks down almost to the level of a mere cradle for the germ-cells, a place in which they are formed, and under favourable conditions are nourished, multiply, and attain to maturity.

immortal germline in his hypothesis on unicellular organisms and in his later hypothesis on germinal selection.

Weismann's germinal selection theory is a hypothesis that outlines possible sources of variation between the individuals of a population which could then allow for natural selection to take place. There could be two types of germinal variation, Weismann argued. The first would be spontaneous germinal selection which in simple, anachronistic terms is similar to later versions of what we now call 'mutation' or 'genetic mutation'. For example, Weismann suggested that the 'congenital deformities of Man may fall in part under this category', using examples such as those born with six fingers.

The second type of variation that Weismann suggested is 'induced germinal' variation. This is less easy to define in relation to a contemporary understanding of evolution. It could be compared to an induced mutation caused by an external effect. However, induced mutation is usually a one-off occurrence caused by extreme abnormal external conditions, such as exposure to radiation. It does not adapt the individual to the 'external influence', and is in most cases considered harmful to the organism. However, Weismann's 'induced germinal selection' allowed for an adaptational response to the environment that was beneficial to the organism, which could then be inherited. Weismann clearly thought that there was a role for induced germinal selection in the formation of new species. However, he was unclear as to the degree to which the direct influence of external conditions affected the germ plasm, and pointed out that this form of variation would always have a subordinate role to personal selection (Weismann's term for natural selection).

So even Weismann, contrary to the way in which he is often depicted, allowed a role for the inheritance of characters that were acquired in response to the environment. It was actually only use and disuse that he sought to discount from evolutionary thought.

Wallace is best known today as Darwin's celebrated 'co-discoverer'. However, he is less recognised for the role that he played in the promotion of 'Darwinism' after Darwin's death in 1882. Wallace was fourteen years Darwin's junior and had the good fortune to outlive his co-discoverer by 31 years, living until 1913. Wallace did not fade into oblivion either – he was one of the most respected naturalists of his age and, partly due to his writing on a multitude of subjects, was to some extent a household name. Even in the early twentieth century he was seen as the last of a group of highly influential men of science of the previous 'wonderful century'. He was referred to in the press as the 'Grand Old Man of Science' or 'The Last of the Great Victorians'. It is from this position that he, perhaps unwittingly, exerted a lasting influence over what we now understand by the term Darwinism.

Wallace's book *Darwinism* was published in 1889. This was one of the most significant British publications in the Darwinian/neo-Darwinian debate of the late 1880s and 1890s. It was in this book that Wallace outlined his version of 'pure' Darwinism – and it is this version of Darwinism that we tend to think of today, one that is wholly reliant on natural selection with no role for the inheritance of acquired characters. Wallace firmly saw himself in the Weismannian camp. Wallace was an ardent supporter of natural selection, attributing far more to the evolutionary mechanism of natural selection than Darwin himself. He was therefore keen to reject any notion of the inheritance of acquired characteristics and effectively wrote Darwinian pangenesis out of his history. It was partly in order to describe Wallace's 'pure Darwinism' that the term neo-Darwinism was originally coined by Darwin's friend

George John Romanes in the 1890s. Romanes and Wallace disagreed on a number of issues, including the origin of the human mind and pangenesis.* In particular Romanes complained that: 'Those biologists who of late years have been led by Weismann to adopt the opinions of Wallace, represent as anti-Darwinian the opinions of other biologists who still adhere to the unadulterated doctrines of Darwin.' He continued: 'So much, then, for the Darwinism of Darwin, as contrasted with the Darwinism of Wallace, or, what is the same thing, of the neo-Darwinian school of Weismann.'

'So what?' one might ask – after all, this seems to be nothing more than a minor footnote in the history of evolution. Indeed, on some levels it is – I am certainly not suggesting that we revisit Darwin's pangenesis or Weismann's induced germinal selection theories, and certainly not 'use and disuse', as models for research today. My point is to highlight that over the last 150 years there has been another strand to the Darwinian history of biology that is often written out of the neo-Darwinist accounts. And there have also been those who fervently believed themselves to be Darwinists, but managed to become seen as anti-Darwinian or, horror of horrors, Lamarckian. This is not just a matter of historical pedantry – nitpicking at some scientist's whiggish and celebratory account of his or her discipline's history – after all, there is plenty to celebrate. The way in which Darwinism has been portrayed to the public throughout the twentieth century has been skewed towards a neo-Darwinian perspective – and the use of historical accounts

---

* While Wallace was a strict selectionist in other respects, he drew the line at the origin of the human mind. The advanced intellectual capacities of Western man were, he felt, the result of a spiritual incursion. Romanes, on the other hand, believed that they were purely the result of a process of evolution by means of natural selection.

has been key to this. But right now this is changing, and quite rapidly so, as scientists are coming to realise that biological systems at all levels are a lot more complicated than previously thought. What was previously seen as unthinkable heresy is currently being openly discussed within the scientific community, and this makes it all the more important to understand the legacy of the way in which Darwinism has been communicated over the years.

Why then is Darwinian pangenesis treated as an embarrassment in some neo-Darwinian accounts of his work? While the idea may be coming back into vogue among some scientists, there are a number of reasons why the wider idea of inheritance of acquired characters has fallen so far from grace that it was until recently seen as a 'no go' area. And some of these reasons are firmly embedded in the politics of the twentieth century.

Stains on 'Lamarckism'

The inheritance of acquired characters (or more often latterly the inheritance of acquired *characteristics*) is a term open to much misinterpretation and derision. It has a convoluted and complex history tainted with scandal, from fraudulent experiments to grotesque political machinations of the highest order. The idea that acquired characteristics could be passed on was damaged among scientists in the West by two now notorious incidents in the interwar years.

First came the sad story told in Arthur Koestler's *The Case of the Midwife Toad*. The story centred on Austrian biologist Paul Kammerer (1880–1926) who experimented on a number of animals, his most infamous experiments involving midwife toads. The midwife toad mates on dry land, and so does not need the mating pads which other toads use to cling on to their mate when mating in water. When Kammerer forced midwife toads to mate in water, they seemed to not only acquire mating pads but pass them on to their offspring – a clear case of the inheritance of acquired characteristics, he argued.

These findings were criticised when they were first published – even by those who firmly supported the idea of inheritance of acquired characters. For instance, in 1923 Joseph Cunningham* wrote that Kammerer's findings were 'in some respects open to the objection that it is not in accordance with the present state of biological knowledge'.

Accusations of fraud had been circulating since 1919, but the final blow came in 1926 when the American biologist G.K. Noble and the Viennese biologist Hans Przibram noticed that the pads on the last surviving specimen had actually been darkened by injecting Indian ink. The exposure of this fraud in the leading journal *Nature* led to Kammerer's subsequent suicide on his way to take up a post in the Soviet Union. Whatever the truth, the damage to Lamarckian inheritance was done.

The Soviet Union was the scene of the second great stain on the Lamarckian character. This was a result of the work of Trofim D. Lysenko. Known as the 'barefoot professor', Lysenko was an agronomist with a peasant background and limited scientific training, who rose to become president of the Lenin Academy of Agricultural Sciences in 1938. Lysenko rejected what he saw as 'quasi-spiritual', bourgeois Western genetics in favour of a 'materialistic', environmentally driven version of the inheritance of acquired characteristics. He promised Stalin he could revolutionise Soviet agriculture with his own brand of botanical engineering.

---

* Joseph T. Cunningham was a marine biologist who sought to reconcile the recently 'rediscovered' Mendelism with the inheritance of acquired characteristics. In 1921, he had published these theories in his book *Hormones and Heredity*.

His most famous idea was to boost yields by planting winter crops – crops planted in autumn for a spring harvest – in spring as well. Winter crops need the cool of winter to stimulate spring growth, so Lysenko chilled germinated seeds to mimic wintry conditions, a process he called vernalisation. He believed that the effects of vernalisation would be passed on. Opponents of Lysenko were silenced, sometimes in the most extreme ways possible.* I have heard, anecdotally, from some of those who were working in the Soviet Union at this time that one could not even say the word 'genetics' publicly for fear of reprisal.

However, it is worth remembering, as J.B.S. Haldane – himself at the time a communist – highlighted in the 1940s, that Lysenkoism was not necessarily a product of the Soviet ideological system; there were also critics of genetics in the West. However, after the darker side of Lysenkoism became apparent, there was a clear and understandable move in the West to cut any ties with this biological regime.

Getting dogmatic

With those two stains on its reputation, and Weismann's assertion of the sanctity of the germline, it is hardly surprising that inheritance of acquired characteristics began to become something of a taboo subject among Western scientists. It received a further hammer-blow in the late 1950s from the intervention of Francis Crick, joint discoverer with James Watson of the double helix structure of the DNA molecule.

In 1958 Francis Crick published a paper entitled 'On Protein Synthesis'. In this paper Crick introduced what he termed the

---

* The most notorious being the imprisonment in 1940 and subsequent death in 1943, reportedly due to malnutrition, of Nikolai Vavilov, the founder and former president of the Lenin Academy of Agricultural Sciences.

'central dogma of molecular biology' for the first time. What Crick said, essentially, was that 'information' could be transferred from DNA, the genetic material in every living cell, to proteins, life's basic materials – that is, DNA could direct the creation of proteins via RNA – but proteins could never affect DNA:

> This [dogma] states that once 'information' has passed into protein it cannot get out again. In more detail, the transfer of information from nucleic acid to nucleic acid, or from nucleic acid to protein may be possible, but transfer from protein to protein, or from protein to nucleic acid is impossible. Information means here the precise determination of sequence, either of bases in the nucleic acid or of amino acid residues in the protein.

If this was so, then it as good as ruled out the possibility that any changes in an organism and its proteins during its lifetime could affect its genetic material. This seemed like a death knell for the inheritance of acquired characteristics, and any hint of Lamarckism became more or less a taboo subject. This has meant that for a good part of the later twentieth century the 'inheritance of acquired characteristics' was seen as the ultimate *Darwinian* heresy. Those who have sought to support or research this area in any way have often been depicted as acting on the maverick fringes of the scientific or academic communities. This is the atmosphere in which Dawkins began his scientific career, and subsequently wrote *The Selfish Gene*.

Understandably, then, Richard Dawkins is perhaps one of this concept's most vehement contemporary critics, seeking to either deny any possibility that this mechanism might occur in nature or to simply negate its relevance if it is shown to occur on any level.

There is a very good reason for this, as there are many who would argue that if any environmentally driven inheritance of acquired characters were to occur in nature – as now seems increasingly likely, given new findings in some areas of research – it would completely undermine Dawkins' 'selfish gene' hypothesis. This is because it could potentially undermine the concept of immortal replicators, or a germline that is shielded from the direct or indirect influence of the environment.

And of course while there are clearly differences between the neo-Darwinism of the nineteenth century and that of Dawkins and his supporters today, there is one underlying theme – an outright rejection of the inheritance of acquired characteristics. So perhaps we ought to see the latter more as a neo-Darwinian heresy than a Darwinian one.

Reading back

But Crick's outlining of the 'central dogma' was not the end of the debate over the inheritance of acquired characteristics. In the 1970s, American geneticist Howard Temin of the University of Wisconsin-Madison, along with Renato Dulbecco and David Baltimore, made a startling discovery about DNA.

The received wisdom was that DNA never makes proteins directly. Instead, sections of it are copied onto strands of RNA, which peel off and use the cell's molecular machinery to build the proteins from amino acids. That way, the DNA stays perfectly intact. So in line with Crick's central dogma, information flows from DNA to RNA to protein – but the reverse never happens. At least, that is what was thought at the time. But Temin discovered how viruses exploit a special enzyme of DNA he called 'reverse transcriptase' that transcribes a single strand of RNA into a double strand of DNA. In this way, it actually reverses the flow of information from RNA to DNA.

Temin's discovery was one of the most important in modern medicine, and has led to an understanding of how many viruses such as HIV and hepatitis B work, as well as underpinning several key methods in molecular biology and diagnostic medicine. In 1975, Temin and his colleagues received the Nobel Prize in Physiology or Medicine for their work. Since then, it has been discovered that reverse transcriptase is common throughout the plant and animal world.

Following Temin's discovery of reverse transcriptase, Crick rearticulated his central dogma to allow for 'special transfers' that occur in unusual circumstances, alongside the normal information transfers – DNA to DNA, DNA to RNA and RNA to protein. He talked about three of these special transfers – RNA to both RNA and DNA in virus-infected cells, as with reverse transcriptase, and from DNA direct to protein in cells taking up the antibiotic known as neomycin.

The end of Weismann's barrier?

None of this, though, was directly implicated in evolutionary theory, even though most neo-Darwinists remain wedded to the central dogma. However, in the 1970s a young Australian immunologist named Ted Steele was struck by an idea. This was that the body's immune system may provide a perfect experimental demonstration of the inheritance of acquired characteristics – the body responding to environmental stimuli then passing its reaction on to offspring. If Steele could demonstrate that an immune response was inherited, he believed, then he would have shown a kind of 'Lamarckian' evolution.

As Steele's fellow Australian Frank Macfarlane Burnet had shown, when foreign material enters the body its marker protein, or antigen, will eventually meet one of the body's immune cells that interacts specifically with that antigen. Immediately they meet, the immune cell produces a flood of clones of itself that swarm around the body latching on to the antigens and attaching antibodies to them, which act as marker beacons for destructive blood cells. Through a process Burnet called 'clonal selection', when an antigen enters the body it acts as a selection agent. It takes on a similar role to that which natural selection plays outside of the organism. Therefore, there is a 'positive' selection of those immune cells which produce the antibodies specifically for the antigen. The result is that there are a larger number of those immune cells in the immune system. Burnet and his colleague Peter Medawar received a Nobel Prize for their work on the immune system in 1960.

Steele thought that Temin's discoveries suggested a way in which genetic information could be sent around the body by RNA viruses, which transcribed the genetic information in one cell and reverse transcribed it to another via reverse transcriptase. With such a flood of identical immune cells reacting to foreign material there is, he suggested, a high chance that their genetic material would be transcribed in this way and infiltrate with the virus into the germ cells, altering their genetic material by reverse transcription. If so, the immune response might be transferred to offspring – and, Steele added, that 'Weismann's barrier' might thus be infiltrated.

Steele and Dawkins

Together with Toronto colleague Reg Gorczynski, Steele conducted experiments in which they injected foreign material into male mice, and observed whether antibodies developed in reaction to it. They claimed to have detected the same immune reaction in a significant proportion of the injected mice's offspring. In 1980, the headlines went around the world that Lamarck had been vindicated and Darwin disproved. Two other teams attempted to replicate Gorczynski and Steele's work. One of these experiments not only replicated it, but also extended it. There were some differences in methodology and the results published in *Nature* in 1981 were open to different interpretations. Some saw this as a refutation of Steele's results – which he contested – and as a consequence there was a minor scandal during which Steele's reputation was publicly shaken.

Nonetheless, Dawkins was aware enough of the implications of Steele's ideas to respond at length in *The Extended Phenotype*. He claimed that even if Steele were proved right it would not bother him because, as far as he was concerned, Steele's theory was Darwinian rather than Lamarckian.

> I now view with equanimity ... the prospect of Steele's theory being upheld, because I now realise that, in the deepest and fullest sense, it is a Darwinian theory.

By Darwinian, of course, he meant neo-Darwinian, because as we have seen, Darwin himself would have had no problem with Steele's own interpretation. Steele did actually see himself as a 'Darwinian' – but in a very different way to the one Dawkins meant. It all comes back to the same disagreement that Romanes and Wallace had over the inclusion of Darwinian pangenesis under the heading of 'Darwinism'. Steele – a pluralist – saw natural selection as one form of evolutionary mechanism like Darwin, and Dawkins – a panselectionist, like Wallace – saw natural selection as the only form of evolutionary mechanism.

Dawkins' argument was that the response of the immune system in Steele and Goczynski's experiments was actually congenial to his own theory of germline replicators, joking that 'Steele's book might be retitled *The Extended Germ-line!*' Steele emphatically refuted this, responding: 'In one breath he [Dawkins] is accepting that a somatic cell *might* contribute to the germ cells, a concept that is the antithesis of Weismann's dogma, and in the next breath he is saying that such a concept is "deeply congenial" to neo-Weismannists.'

Steele was considered something of a maverick by some evolutionary scientists, and his 1998 book *Lamarck's Signature* received a vicious pasting in a *New Scientist* review by evolutionary scientist Professor D. Laurence Hurst, who summed up Lamarck's theory as follows: 'if you chop off a rat's tail, its offspring will also have no tails', and said that Steele was making 'a stupefyingly large mountain out of a hypothetical molehill'. Dawkins was more diplomatic but insisted that even if Steele's argument that acquired immune responses can be passed on is correct, it is still within the neo-Darwinian framework and is not, as Steele had suggested, proof of the reality of Lamarckian acquired inheritance. A lot of this debate really comes down to semantics and how one chooses to define the term 'inheritance of acquired characteristics' – close analysis of exchanges between the two reveals that Steele and Dawkins hold to very different ideas of what the term represents. Steele predominantly used it to describe the effect of the external environment on the soma and germline, and Dawkins tended to refer to use inheritance. As we have discussed, these are very different mechanisms.

Gene bypass

Yet if Steele's attempt to breach the Weismann barrier and knock down the central dogma has so far failed to really disconcert dyed-in-the-wool neo-Darwinists, there is now a threat to the citadel from another angle: the developing science of epigenetics. Steele tried to show how changes during an organism's lifetime could be inherited through the germline, but epigenetics may be showing how DNA may be bypassed altogether, enabling traits to be passed on 'epi-genetically' – that is, outside the genes. If this is the case, then the gene-centric view of many neo-Darwinists might start to look rather outdated.

The phrase was originally coined in the 1940s by the biologist Conrad H. Waddington to describe how genes might interact with their environment to produce the form of the organism, its phenotype.

But the focus and meaning of epigenetics have both changed substantially. Modern biologists disagree about precisely what it does mean, in much the same way that there was no concrete agreement about what 'acquired characters' meant in the nineteenth century, but the core of the science is how genes are 'expressed' – that is, just what effect the genes have on the organism. Epigeneticists talk about how particular genes may be 'marked', or switched on and off.

Genes are the organism's instruction book, but they have to be told what to do and when to do it. When an organism is first conceived, for instance, all of its cells are identical and can develop into any other kind of cell. But as the organism develops, cells differentiate – that is, they become different, with some becoming nerve cells, some liver cells, some blood cells and so on. Every cell might contain identical DNA, but a liver cell knows how to be a liver cell, and a blood cell knows how to be a blood cell, because an array of chemical markers, which are today known as the organism's 'epigenome', turn on and off particular genes in its DNA. If DNA is analogous to ROM, the cell's fixed memory, the epigenome is the RAM, running programs and responding to inputs.

Nurture switches nature

Most work on epigenetics is concerned with how genes are expressed during an organism's lifetime. Here it is becoming increasingly clear that the environment really does play a part. In other words, environment can influence the expression of genes. Diet, the presence of toxins, and temperature can all have an effect on how genes are expressed, turning them off or switching them on. This discovery is opening up new avenues of research that could have major implications for human health. One early bit of research by Ming Zhu Fang and colleagues at Rutgers University, for example, suggests that green tea might epigenetically help to prevent cancer-fighting genes in the body from being switched off, and so could prevent cancers. It also opens up the possibility of treating disease epigenetically. Already one epigenetic drug, 5-azacytidine, has been approved by the Food and Drug Administration in the US for use against myelodysplastic syndrome, also known as preleukemia or smouldering leukemia. At least eight other epigenetic drugs are currently in different stages of development or human trials.

There have been some fascinating hints that even something as intangible as the quality of mothering could have an effect on the expression of genes. In an experiment by Michael Meaney at McGill University, mice that were licked by their mothers when young grew up brave and calm because the licking switched on certain genes; those mice that were not licked properly grew up neurotic as the same genes were left blocked. Meaney is now embarking on a major research programme to look at the epigenetics of human nurturing, examining, for instance, if a depressed mother who finds it difficult to bond with her baby might alter the baby's brain epigenetically. There may be a more positive side; for just as genes can be switched on epigenetically, so they could be switched off epigenetically. And so it is just possible that we could find ways to alter the gene program by such simple things as changes in diet.

The realisation that every cell carries epigenetic markers that influence the expression of its genes has opened up something of a Pandora's box. It is clear that the human genome, with its three billion base pairs that were mapped with such painstaking effort by the Human Genome Project, is just the tip of the iceberg. Microbiologists are now tentatively embarking on a Human Epigenome Project. But it is a massive task, with every kind of cell carrying its own epigenome – and every epigenome potentially changing throughout each person's life.

The ghost in your genes

It is now hard to doubt that the environment has a powerful impact on genes during an organism's lifetime through epigenetics. What is also emerging is that the impact of the environment can also be passed on to offspring epigenetically. Botanists have known about transgenerational epigenetic effects in plants and fungi for some time. Now it is beginning to emerge that they happen in animals, too.

One well-known example, noticed by Ralph Tollrian in 1995, is that of daphnia water fleas, which develop protective spines in the presence of predators and pass them on to their offspring. But even mammals show these effects. In a key experiment in 2000, Randy Jirtle and Robert Waterland of Duke University conducted an experiment with frail, yellow, obese cancer-prone mice called agouti mice, who suffer in this way because of a gene called the agouti gene. Amazingly, by feeding the mother agouti mice a diet rich in chemicals called methyl-donors – found in such foods as garlic and onions – they managed to switch off the agouti gene in their babies, so that they were born normal, brown and healthy. This study and others have thrown up the possibility that a baby's epigenetic profile may be affected by what the mother eats when she is pregnant.

Even more fascinating transgenerational epigenetic effects in humans have been revealed by historical studies. Looking at crop data and population from the isolated village of Överkalix in northern Sweden over two centuries, Marcus Pembrey and Lars Olov Bygren discovered that the quality of teenagers' diets affected the health of their grandchildren. Teenage boys brought up in times of famine, for instance, had grandsons that lived longer, while teenage boys raised in times of plenty tended to have grandsons who became diabetic. Interestingly, the effects were sex-specific, so that girls brought up in times of abundance had grandchildren who were less prone to heart disease.

The new Lamarckists

Some biologists, such as Eva Jablonka of Tel Aviv University, believe that epigenetic effects are proof that perhaps Darwin and Lamarck were right about the role of the environment after all. According to Jablonka and Marion Lamb:

> In evolutionary studies, because heritable non-genetic variations are often induced by the environment, we have to expand our notion of heredity and variation to include the inheritance of acquired variations, the once disparaged

idea that was part of Lamarck's theory. In a sense, we have to go back to Darwin's original, pluralistic convictions. Darwin, unlike many of his more dogmatic followers, saw a role for induced variation in evolution. Today, in the light of the newly discovered epigenetic mechanisms, Darwinian evolution should include descent with epigenetic as well as genetic modifications, and natural selection of induced as well as random variations. Certainly, it should not be reduced to 'selfish genes.'

Others argue that epigenetics is a 'red herring' as far as Lamarckian inheritance is concerned, because it does not involve rewriting genes; it merely alters gene expression in offspring while leaving the genes intact. Epigenetic inheritance is apparently easily revers-ible, and there is as yet little evidence that it persists for more than a few generations. But, just as genes can mutate and be selected for, epigenetic effects, such as DNA methylation patterns, can mutate and be selected for and so still may play an important role in evolution.

However, there is no doubt that the research into epigenetic mechanisms will come to play a powerful role in our understand-ing of evolution. Even if it does not turn out to prove Lamarck (and Darwin instead of Dawkins) right, it is already showing itself to be a valuable survival tool for organisms, allowing them to adapt within their lifetimes to changes in their environment. It may be that it will prove to be another key evolutionary mecha-nism, working in tandem with mutation and natural selection. Whatever the truth turns out to be, it is clear that Darwin's more pluralistic view of evolution may have more of a future than the narrow confines of the selfish gene hypothesis.

CHAPTER 5

LIFE STORIES

These facts … are leading us in the direction of a central truth about life on Earth … This is that living organisms exist for the benefit of DNA rather than the other way around … Each individual organism should be seen as a temporary vehicle, in which DNA messages spend a tiny fraction of their geological lifetimes.

Richard Dawkins, *The Blind Watchmaker* (1986)

There is no doubt that Richard Dawkins believes passionately that he has seen the central truth about life on earth, and wants to spread it to the world. The very certainty and clarity with which he expresses his ideas is attractive for readers too. It is so easy to be swept along by the power of his rhetoric and the grand simplicity of his vision. The idea that life on earth in all its startling variety and dynamism can be explained entirely by the self-perpetuating 'selfish' nature of strands of DNA alone is appealing both for its logic and its utilitarianism.

But just because an idea is powerful and simple, this does not mean that it is necessarily 100 per cent right. Sometimes, reality may be just a little more messy and complicated than that, which is why scientists not only keep testing their ideas against reality but also make objective observations, too – however potent the logic of their theories.

Dawkins is undoubtedly a master of argument, and the selfish gene concept is remarkable in its ability to adapt to all kinds of challenges and new ideas. But that does not mean that it must be entirely right. 'Nothing is more dangerous than a dogmatic worldview', said Dawkins' most weighty critic Stephen Jay Gould, 'nothing more constraining, more blinding to innovation, more destructive of openness to novelty.' Dawkins would surely agree, and yet there is a danger of Dawkins' view of evolution becoming just that, dogmatic. It is not that he is inflexible. He is amazingly ingenious in his ability and willingness to reshape his arguments to accommodate challenges, as we saw with Ted Steele's ideas. But science is not always about ingenuity, however beguiling; it is about listening to the evidence – and the evidence now coming in from biologists around the world is that the 'selfish gene' account of natural selection may not be able to provide a complete picture of evolution.

Going for Gould

It is interesting that the late Stephen Jay Gould and his supporters provided one of the longest-running and highest-profile challenges to the Dawkins hegemony. Gould was a palaeontologist, a field scientist who spent his entire life – when not writing books – looking at fossils, and trying to understand what he saw in the fossil record. This is not to say that there is no place for occasional armchair theorising, or even that Gould was necessarily any more correct than Dawkins. He could certainly be equally uncompromising and vituperative in his arguments. But I would suggest that when someone who spends a lot of time directly examining the evidence says that our theories are wrong or need adjusting, then it may be worth at least listening to them.

The Gould-Dawkins battle reached a climax when Gould reviewed a book by the Oxford philosopher Daniel Dennett, one of Dawkins' staunchest allies, for the *New York Times*. The book was *Darwin's Dangerous Idea* – in it Dennett championed

Dawkins' version of Darwinism, which he saw as a universal acid, capable of corroding away all of the unhelpful intellectual residue of the past. Dennett was particularly scathing about Gould's work, bracketing it with 'Intelligent Design' as a 'skyhook' – an argument to explain complexity that is not built up from solid, simple bases like a crane but simply dangles from a sky of nothingness. In his review Gould responded, saying:

> Daniel Dennett devotes the longest chapter in *Darwin's Dangerous Idea* to an excoriating caricature of my ideas, all in order to bolster his defense of Darwinian fundamentalism. If an argued case can be discerned at all amid the slurs and sneers, it would have to be described as an effort to claim that I have, thanks to some literary skill, tried to raise a few piddling, insignificant, and basically conventional ideas to 'revolutionary' status, challenging what he takes to be the true Darwinian scripture. Dennett claims that I have promulgated three 'false alarms' as supposed revolutions against the version of Darwinism that he and his fellow defenders of evolutionary orthodoxy continue to espouse … Since Dennett shows so little understanding of evolutionary theory beyond natural selection, his critique of my work amounts to little more than sniping at false targets of his own construction. He never deals with my ideas as such, but proceeds by hint, innuendo, false attribution, and error.

Some have suggested that the heat generated between Gould's camp and that of Dawkins was as much about territory and status as ideas – or at least that it was between Gould and Dawkins' fundamentally different stances on the relationship between science and religion, which I shall explore further in the next chapter. However, it does seem clear that there was also a fundamental disagreement on the science.

Replicators and species

For Dawkins, life is fundamentally about replicators. Ultimately nothing else matters in the story of life but these, which are usually the microscopically small knots of chemicals we call genes, but can also be the cultural replicators that Dawkins calls memes. Life forms, from the lowliest bacteria to human beings and other mammals, are no more than 'vehicles' built by genes, which have combined in order to improve their chances of survival. Genes, according to Dawkins, do not always work together; they can behave as outlaws, sabotaging their vehicles for their own benefit. They can also have 'extended phenotype' effects on the environment beyond their immediate vehicle, such as a beaver's dam. According to this view the development of life through evolution is entirely about the natural selection of genes; random mutations that do not improve survival are eliminated, and the few that do prosper. Through gradual changes in genes, driven by natural selection, populations or lineages become adapted to their physical circumstances.

For Gould, on the other hand, evolution was essentially about species. Looking at the fossil record, what he saw was species, not genes. It was not simply that concrete examples of species are preserved as fossils and genes are not; Gould believed that the fossil record showed that groups of organisms were the level at which selection occurred, not genes. Moreover, he argued that it happened in a very different way from the picture painted by Dawkins and the neo-Darwinists.

The conservatism of life

According to Gould, the neo-Darwinist view envisaged a gradual process of improving adaptation, and increasing complexity and variety, through natural selection over the entire span of life on earth – frequently interrupted by catastrophes such as meteorite impacts, but essentially moving on all the time. Gould, however, believed that the fossil record shows something very different. The extraordinary and unique preservation of the soft tissues of sea creatures from some 505 million years ago in the Burgess Shale* seems to show an extraordinary range of different phyla compared to what we have now. Gould argued that this reduction in the numbers of phyla has led to a reduced range of body plans currently available for lifeforms to take and that life, while becoming more diverse in terms of number of species, has actually become less varied in its range of basic forms since the first explosion of multicellular life, of which the Burgess Shale find forms a record. There may be well over 350,000 different kinds of beetle, for instance, but they are all beetles – with the same number of legs, the same body plan and much more besides.

Moreover, Gould and his erstwhile colleague Niles Eldredge argued, the fossil record shows that species change little from the first time they appear in the record to the time they disappear. Evolution, it seemed to them, jumps from one species to another, with long periods of stability in between – in complete contrast to

---

* Discovered in 1909, the Burgess Shale fossil bed in the Canadian Rocky Mountains provides the best record we have of Cambrian animal fossils from 505 million years ago. The Cambrian period is of particular interest as it followed what is referred to as the 'Cambrian explosion', a period 530 million years ago when an incredibly diverse number of species suddenly start to appear in the fossil records. However, the term 'Cambrian explosion' is sometimes disputed, as it is argued that some of the animal species thought to have appeared during this period may date from much earlier.

the gradual but continuous change of the neo-Darwinian model as it is usually understood. Gould and Eldredge therefore came up with a model of evolution which they called 'punctuated equilibrium', which envisaged long periods of stability during which species changed little, occasionally interrupted by sudden shifts to a new species or range of species.

Creeps or jerks?

Punctuated equilibrium was at first widely criticised because it seemed to imply a series of sudden massive mutations, like an animal growing wings overnight. But Gould and Eldredge clarified that when they were talking about sudden shifts, they meant sudden only in geological terms. A species that lasted for 2 million years, for instance, might come into existence over just 50,000 years, which is a mere instant in geological terms. But it is plenty of time for thousands of generations of a species to live, reproduce and die in.

Responding to this idea in his subsequent book *The Blind Watchmaker*, Dawkins suggested that Gould and Eldredge were caricaturing the neo-Darwinist view of the continual progress of evolution as 'constant speedism'. He countered that the picture that he and his allies were really presenting was of 'variable speedism' in which the rate of evolution continually changed, sometimes speeding up, sometimes slowing down. And so, Dawkins argued, Gould and Eldredge's periods of stability were just extreme slow periods while their sudden shifts were extreme fast periods – but there was no sense of interruption. Punctuated equilibrium is, Dawkins argued, a 'minor gloss', an 'interesting but minor wrinkle on the surface of neo-Darwinian theory', and 'lies firmly within the neo-Darwinian synthesis'.

Limits to evolution

Yet if this was so, why did Gould and Eldredge deny it so firmly and Dawkins equally vehemently attack Gould and Eldredge? My feeling is that it was because punctuated equilibrium reduced the role of natural selection, especially at the gene level. Gould was an ardent Darwinist but he argued, unlike the neo-Darwinists, that the fossil record implies that there is more to evolution than natural selection.

First, Gould suggested that the range of possibilities is limited by some mechanism as yet unknown – and that that is why evolution has been so conservative over such long periods. There is good evidence that nearly all the 30-odd phyla of the present-day animal kingdom emerged hundreds of millions of years ago – and that no new ones have emerged recently. So if evolution has not exactly stopped, it at least seems to have slowed down the development of new phyla. It seemed to Gould, then, that the variation produced by natural selection must be limited.

Second, the fossil record shows that, from time to time throughout the long history of the earth, huge numbers of species disappear and new ones appear in events called 'mass extinctions'. The most famous, of course, is the one that wiped out the dinosaurs. Gould argued that such events played a huge part in the evolution of new species, but the process he described was very different from the versions of gradual adaptation solely by means of natural selection in the neo-Darwinian mode. The dinosaurs did not die out, Gould argued, because they were not well adapted; they were well enough adapted to have dominated the earth for 160 million years. They died out because they were unlucky enough to have been around when a meteorite struck the earth. No amount of adaptation could have prepared them for that.

Going out with a bang

Mass extinction events have certainly culled huge numbers of animals. According to Gould those that survived did so because they already had traits that allowed them to weather the disaster and cope with the new conditions. So they were naturally selected; however, it was not a gradually acquired adaptation that allowed them to survive, but qualities they happened to have when the conditions began to change. This, Gould argued, means that selection happens at the level of *species*, not *genes*, and works more by chance than by gradual adaptation. Gould suggested, for example, that the fossil record really showed that the classic evolutionary story of the horse, evolving from dog-sized forest dweller to the large grassland animal of today, is not a story of gradual adaptation and increasing diversity, but one of extinction – all the other varieties of horse in the past died out as their habitats were lost, leaving only the modern horse as a survivor.

In his book *Wonderful Life*, Gould introduced the famous metaphor of 'replaying the tape' of life's history. Because the chance events that cause mass extinctions play such a key role, Gould believed that if the whole 'tape' of natural history were to be played again from the beginning, there is no reason to expect that life would turn out again the same way as it is today, as these chance events would not happen in the same way. The message of this is that sudden changes in environmental factors play a key role in the way that evolution unfolds. Under Dawkins' adaptionist model of evolution, the environment poses challenges which gradually, over time, act a kind of sieve that individual variations or replicators must pass through in order to survive. Under Gould's model the environment can have a sudden catastrophic impact on species, acting to favour species over one another, rather than individuals.

Interestingly, a recent long-running experiment by Richard Lenski at the University of Michigan has produced some evidence of what seem to be 'sudden' beneficial mutations. In 1988, Lenski started growing twelve identical populations of the bacteria *Escherichia coli*, fed them on glucose and watched them slowly evolve over twenty years through 40,000 generations. Most of the populations evolved as expected, growing bigger and better able to consume glucose. Yet suddenly, after 31,500 generations, the bacteria in one of the populations started feeding on citrate in the food, which *E. coli* cannot normally metabolise. It was as if honey bees had suddenly acquired a liking for vodka. The apparently 'sudden' beneficial mutations appeared to be dependent on the bacteria having had previous non-beneficial 'potentiating' mutations already. It is not yet quite clear what this means, but the research was certainly a blow for Intelligent Design supporters who insisted that dramatic changes cannot happen 'randomly'. As a result of this, Lenski received sustained attacks on his credibility from some pseudo-science establishments or those seeking to support intelligent design.

Sliding genes

There is another new area of research – into the possibility of horizontal gene transfer – that is currently throwing up interesting new challenges to old ways of thinking about evolution. Horizontal gene transfer is the transmission of genetic elements not down the generations through the process of reproduction, but sideways between individuals of the same generation. Most interestingly, this kind of transfer can happen between organisms which are members of vastly different species. These new genetic elements can then be passed down the generations via the usual process of reproduction.

In the 1990s, studies of bacteria began to show that horizontal gene transfer is rife at the microbial level, with genes jumping between bacterial genomes with startling promiscuity. These discoveries immediately became a hot potato, as biologists such as W. Ford Doolittle suggested in 1999 that they showed that 'the history of life cannot properly be represented as a tree'. This was a pretty striking claim, since the tree of life is central to Darwinian thinking.

It would be fair to say that Darwin's ideas had their moment of birth when, as a young man in July 1837, he had a flash of inspiration and drew a little sketch of a tree to represent the evolution of life, with a modest little note saying 'I think' next to it. Since then, Darwin's spindly bush has grown into the huge oak at the core of Darwinist thought. The idea that all life has descended from common ancestors, spreading and branching out into diverse species through time, is in some ways an essential part of the neo-Darwinian view of evolution. Evolutionary biologists have spent lifetimes of research trying to create a complete tree that shows the evolution of life, looking for common ancestors and tracing lines of descent. But this model depends on genes being passed vertically from generation to generation. If they can also be swapped horizontally, then the tree begins to look decidedly shaky.

This is an area of research that has gained significant ground in the past few years. It has become more and more apparent that due to horizontal gene transfer the relationships between species in groups such as the prokaryotes (like bacteria) and unicellular eukaryotes (like slime moulds and algae) is fluid, to say the least. There is also growing evidence that this process could have massive implications for the way in which we understand the origins of plant and animal species.

This subject recently exploded onto the public stage. The 24 January 2009 edition of *New Scientist* was provocatively titled 'Darwin Was Wrong: cutting down the tree of life'. This title – which understandably elicited a strong response from some members of the scientific community – referred to an article on horizontal gene transfer.

Dawkins' response was emphatic – he commented in a recent public lecture that 'already all over the United States creationists are waving this copy of *New Scientist* as if triumphantly, they don't bother to look inside the magazine, as the Editor must surely have known when he sanctioned that headline, a disgraceful piece of publicity seeking'.

However, there comes a point at which we just have to accept that there will always be a minority who believe in a 'creationist' account of the origin of man, in much the same way as there will always be people who think that they have been abducted by aliens. We do not refrain from talking about the possibility of bacterial life on Mars because it might give fuel to alien abduction theorists, so why should we not publicly debate certain aspects of the exciting new developments in evolutionary biology? It would be a great shame if the threat of creationism were to discourage us from debating any relevant challenges to the current 'Darwinian' model of evolution, lest we become dogmatic ourselves.

Creationists will find fuel in whatever we say – a typical tactic is to paraphrase from books out of context and change the meaning of what is being said. No doubt some creationists might try to use this very book to support their stance – but is that a valid reason not to write it? Science should not be about an adherence to dogma born from the 'threat' of creationism. But just to be absolutely clear – I think creationism, in the words of Sir Edward Fry, is both 'false science' and 'false theology'.

Shaking the tree of life?

It's no wonder, given these tensions, that some biologists have fiercely challenged the significance of horizontal gene transfer. In 2006, a team from the European Molecular Biology Laboratory in Heidelberg, led by Peer Bork, examined 191 genome sequences from a wide range of organisms, from *E. coli* to elephants. They identified 31 genes that were common to all of these species and showed no signs of ever being transferred horizontally. These core genes seemed to provide strong evidence for common descent. But other biologists challenged the significance of the research, suggesting that 31 genes is a vanishingly small proportion of the total number of genes in even the simplest genome, which usually runs to tens of thousands.

The classic example of horizontal gene transfer which has long been accepted, and is probably already familiar to most people, is the transfer of antibiotic resistance. Bacteria can pass transposable elements, or genetic information that gives them immunity to a type of antibiotics, between one another. The bacteria don't even have to be of the same species – the immunity can be passed to different 'species' present in the same microbial community. This goes some way towards explaining why we have such a problem with antibiotic resistance. Under the old understanding of linear transfer of genetic information the spread of resistance – even in bacterial communities which reproduce at a relatively rapid rate – would be a lot slower than we now realise it to be. Horizontal gene transfer means that it is not just one strain that can pass on immunity to the next generation, but any bacteria in the same community.

At first it seemed as if horizontal gene transfer affected only single-celled microbes, which reproduce asexually. Yet there is increasing evidence that it occurs in animals and plants too.

According to the aforementioned recent article in *New Scientist*, horizontal gene transfer 'has been documented in insects, fish and plants, and a few years ago a piece of snake DNA was found in cows. The most likely agents of this genetic shuffling are viruses, which constantly cut and paste DNA from one genome into another, often across great taxonomic distances. In fact, by some reckonings, 40 to 50 per cent of the human genome consists of DNA imported horizontally by viruses, some of which has taken on vital biological functions.'

Horizontal gene transfer may not completely bring down Darwin's tree of life, but it almost certainly raises some questions about what is happening at the tree's roots, and some see it as a challenge to the idea of a linear and immortal, or insulated, germ-line. However, it is important to note that horizontal gene transfer does not necessarily challenge Dawkins' selfish gene metaphor – after all, these transposable elements could be seen as a sort of selfish replicator as outlined in *The Extended Phenotype*. After all, it is just another way for 'genes' to replicate themselves, and genes only have to have the *potential* to be immortal to fit Dawkins' definition of 'selfish'. It does, however, raise the question – how far can a metaphor stretch before it snaps?

In light of this fascinating new area of research, the selfish gene concept is helpful only up to a point. With the increasingly fluid picture of the relationships between species, it really only reveals a small part of a much bigger area. As philosopher of biology John Dupré commented in the offending *New Scientist* article: 'Our standard model of evolution is under enormous pressure. We're clearly going to see evolution as much more about merger and collaboration than change within isolated lineages.'

Group practices

Other cracks in the standard model of evolution have begun to appear over the question of selection levels. Dawkins has insisted that the only level that really matters is that of genes or replicators. Neo-Darwinists explained altruism, such as the apparently selfless behaviour of worker bees in toiling entirely for the good of the colony, as being caused by selfish genes working to produce 'kin selection' – that is, helping a close relative in order to ensure passing on one's own genes. But they baulk at the idea of group selection.

One of the reasons this matters is because it affects explanations of altruism – a core element of the selfish gene concept. A group, like an ant colony, that co-operates may have a better chance of survival than one in which the members compete entirely for their own needs. But how would these organisms identify who they should be behaving altruistically towards – i.e. those with the same genes as them – if they are not direct kin? The problem for the selfish gene idea is to explain how such a group could evolve, since it would be difficult for the vehicle organisms of altruistic behaviour genes to identify other vehicles of the same genes. Dawkins does attempt to counter this by suggesting that organisms may have perceptible traits – the metaphorical example he gives is a green beard – which help to identify the other carriers of the altruistic trait gene. However, such a group is deeply vulnerable to the 'selfish' interloper who develops a 'green beard' without the altruistic behaviour. So group selection would imply an entirely different mechanism from the selfish gene.

A weird infatuation

As Dawkins has said, 'The interesting question is whether any adaptation of a wild animal or plant is interpretable as group selection. I don't think it is.' Dawkins chose his words carefully when he said 'any wild animal or plant', because group selection has already been clearly noted in colonies of bacteria that form 'biofilms', in which some bacteria 'selflessly' form a mat for others to thrive on. Some biologists are now looking for signs of it in higher organisms too, and are confident they will find it, although just how significant this would be remains to be seen.

Interestingly, it is over group selection that Dawkins recently had another major spat with two eminent American theorists, the Harvard professor Edward O. Wilson and Professor David Sloan Wilson (not direct kin). The idea of group selection in social insects like ants was thought to have been killed off entirely by the neo-Darwinists in the 1960s with their kin selection theory. But E.O. Wilson and D.S. Wilson revived the debate in 2007. In a response in *New Scientist*, Dawkins argued that the Wilsons had simply failed to understand how kin selection works: 'He falls for the first of my "12 misunderstandings of kin selection"; that is, he thinks it is a special, complex kind of natural selection, which it is not', and in a swipe at E.O. Wilson's 1975 classic on sociobiology *Sociobiology: The New Synthesis* went on to say, 'Evidently Wilson's weird infatuation with "group selection" goes way back; unfortunate in a biologist who is so justly influential.' Wilson replied, with some dignity and a certain amount of hubris: 'I am used to taking the heat, and in the past I turned out to be right.'

Toolkits for life

One more area where questions are being asked about the standard model of evolution is in the study of the evolution of the way organisms develop through their lives. Darwin derived key support for his theories from the fact that embryos seem to develop through stages in a way that bears some similarities to how descent from a common ancestor might look. This idea gained a whole new lease of life as 'evo-devo' (short for evolutionary developmental biology) with the discovery of 'toolkit' genes in the 1980s.

Toolkit genes are genes that code not for proteins to be used in the body as everyday 'housekeeping' genes do, but for the development of an organism, switching other genes on and off. There was great excitement in the 1980s over the discovery of homeobox, or Hox, genes in fruit flies. Hox genes seem to determine where limbs and other body segments develop in the fly's embryo or larva. Mutations in these genes can turn a fly's antennae into legs. This could work quite well under the selfish gene approach. The idea of Hox genes as 'master controllers' plays into the idea of genes having long, impressive reaches, determining all sorts of things in the 'vehicle'. They have been conserved, according to this account, because any mutation in them is disastrous for the organism, and so is selected out. An argument against this point is that these genes, just like all others, are tightly regulated by other genes, hormones, cell-to-cell signalling and so on, making their effects context-dependent (i.e. dependent on the developmental milieu).

Evo-devo

At first sight, these genes seemed to be the answer to one of the basic questions about life – how different organisms grow into particular shapes, and how the identical cells of the embryo diverge into all the different forms in the right places to create a multicellular organism. And they have gone a little way down the path of explaining this, which is why evo-devo has become such a huge and growing area of research. Studies are throwing up more and more instances where the same toolkit genes have produced a similar result in organisms separated by millions, if not hundreds of millions, of years of divergent evolution. Toolkit genes that control the creation of contractile muscles that become a heart seem to be shared by mammals, fish and even flies. It has also thrown up many surprises. Darwin was puzzled, for instance, by how some cave-dwelling animals lost their ability to see. It was not an evolutionary advantage to lose the use of their eyes, so he wondered how it could happen; his eventual answer came close to Lamarckism in that he attributed their blindness 'wholly to disuse'. In fact, it seems that the blindness results from toolkit genes which actively 'kill' the eyes during development.

What evo-devo is showing is that the standard picture of evolution, in which natural selection weeds out any traits in an organism which are not helpful and encourages those that are, is inadequate. Biologists are beginning to see that the process is much more complex than that. Toolkit genes and development are both subject to an elaborate mesh of nonlinear interactions and feedback mechanisms that make it seem as if an organism develops in constant conversation with itself. There is some evidence that organisms actively influence their own evolution through some proteins. There is also some evidence that evolution can be sudden rather than gradual because of feedback loops. You only have to throw one stone in a pond to create a whole series of waves, and it may be that new features can develop quite suddenly in a similar way. Stuart Newman of New York Medical College recently showed how limb cells in embryo chickens encouraged other limb cells to grow cohesively together as they developed. If so, Gould and Eldredge's picture of sudden spurts of evolution could be more accurate than the 'variable speedism' of Dawkins.

All these developments are showing that evolution may be far, far more complicated than the neo-Darwinist synthesis allows. It may be that the idea of evolution solely or predominantly by means of natural selection has to be rethought. It may be that the whole idea of a tree of life has to be revised. It may be that the selfish gene metaphor turns out to be quite limited in its reach, and that there are other processes by which natural selection proceeds. My contention at this point is, though, that we should not treat these challenges to the present orthodoxy as challenges to science, fit only to be shouted down or squeezed into an already creaking metaphor, as Dawkins sometimes does. Rather we should see them as causes for excitement, to be embraced as science delves into the real, messy and wonderfully complex world of life.

PART II

PART II

CHAPTER 6

THE CHURCH OF DAWKINS

*I see no good reason why the views given in this volume should
shock the religious feelings of any one. A celebrated author and
divine has written to me that 'he has gradually learnt to see
that it is just as noble a conception of the Deity to believe that
He created a few original forms capable of self-development
into other and needful forms, as to believe that He required a
fresh act of creation to supply the voids caused by the action of
His laws'.*

Charles Darwin, *On the Origin of Species*,
second edition (1860)

As so often, Darwin is worth listening to. Evolutionary theory
is not necessarily incompatible with religious belief, and there
is no automatic need for anyone of religious faith to be shocked
by it.* Yet Dawkins would disagree. He seems to want religious

---

* With a little help from the *Oxford English Dictionary* (direct quotes are in
italics below), here is a summary of some of the different religious takes on
Darwinism:

Creationism – *The theory which attributes the origin of matter, the different species of animals and plants, etc., to 'special creation'.* The important thing to note
here is that each individual species would under this model be made individu-
ally by means of special (divine) creation. In this view there is no evolutionary
process at all – especially when it comes to the soul. Note that this term was not
in use until the mid-nineteenth century.

Young earth creationism – As above, but also rejecting the idea of geological
time. A literalist interpretation of religious texts that suggest the earth is no
more than 10,000 years old.

faith to be crushed by the weight and power of natural selection. A true understanding of evolution must, Dawkins appears to believe, provoke the same loss of faith that afflicted Darwin, and which apparently made him an atheist. In his view, if the theory of evolution does not make you an atheist, you probably have not really got the message. 'Understanding Darwin [does not] drive you inevitably to atheism', Dawkins recently said, 'but it certainly constitutes a giant step in that direction.' Yet it is not altogether clear whether or not Darwin himself was an atheist at all.

Doubting Darwin

In 1860, Darwin wrote to the American botanist Asa Gray, 'With respect to the theological view of the question: This is always painful to me. I am bewildered. I had no intention to write atheistically,

---

Theistic evolution – An allegorical interpretation of religious texts which allows for evolution. Note that this is the mainstream perspective; for example, it is the stance adopted by the Anglican and Catholic churches. It is important to note here that allegorical interpretations of Genesis vastly pre-date nineteenth-century evolutionary thought.

Theism – Generally used to mean belief in a deity or in deities. In contrast to deism (and polytheism), it also means *belief in one God as creator and supreme ruler of the universe, without denial of revelation.*

Deism – *Belief in the existence of a Supreme Being as the source of finite existence, with rejection of revelation and the supernatural doctrines.* Deist – *One who acknowledges the existence of a God upon the testimony of reason, but rejects revealed religion.* Note that deism and allegorical theism were to some extent interchangeable up until the end of the seventeenth century.

Agnosticism – The view of *one who holds that the existence of anything beyond and behind material phenomena is unknown and (so far as can be judged) unknowable, and especially that a First Cause and an unseen world are subjects of which we know nothing.* This term was first coined in 1868 by Darwin's friend and supporter T.H. Huxley, who was known as 'Darwin's bulldog'. It is important to note that Huxley applied this term to himself.

but I own that I cannot see as plainly as others do, and as I should wish to do, evidence of design and beneficence on all sides of us.'

Darwin's own relationship with his faith gradually changed throughout his life. In 1876, in his *Autobiography* which was written for his children and not necessarily for publication, Darwin recalled that at the time of writing *On the Origin of Species* he had wondered at 'the extreme difficulty or rather impossibility of conceiving this immense and wonderful universe, including man with his capacity of looking far backwards and far into futurity, as the result of blind chance or necessity. When thus reflecting I feel compelled to look to a First Cause having an intelligent mind in some degree analogous to that of man; and I deserve to be called a Theist.' However, he went on to say that this conclusion had gradually grown weaker and fluctuated throughout his life, and 'then arises the doubt — can the mind of man, which has, as I fully believe, been developed from a mind as low as that possessed by the lowest animal, be trusted when it draws such grand conclusions?' He later continued: 'The mystery of the beginning of all things is insoluble by us; & I for one must be content to remain an Agnostic.'

But it is the following quote, in a letter from Darwin to a Mr J. Fordyce in 1879, that his son Francis felt most adequately summed up his father's beliefs:

What my own views may be is a question of no consequence to any one but myself. But, as you ask, I may state that my judgment often fluctuates ... In my most extreme fluctuations I have never been an Atheist in the sense of denying the existence of a God. I think that generally (and more and more as I grow older), but not always, that an

Agnostic would be the more correct description of my state of mind.

Of course, we can never know what Darwin's private unwritten thoughts were, thoughts he may not even have shared with his earnestly Christian wife Emma or his children. But most historians conclude that he was an agnostic at most. It is therefore worrying that Darwin and his theories are sometimes appropriated today to support the anti-religious agendas of radical atheists like Dawkins. Does it really matter, though, whether or not Darwin was an agnostic? The important point, surely, is that he clearly did not see his work as *atheistic* nor did he set out to publish *On the Origin of Species* with the intention of advancing atheistic thought.

Dawkins, however, has been known to insist, in a spectacular display of whiggishness, that this is not the case. In his reply on his website to an article in the *Guardian* by Madeleine Bunting (published 28 December 2008), in which he refers to her as 'an ignoramus', he suggests that whatever Darwin himself said about it, he himself knows better: 'It is true that Darwin declined to call himself an atheist. But his motive, clearly expressed to the atheist intellectual Edward Aveling (incidentally the common-law husband of Karl Marx's daughter) was that Darwin didn't want to upset people. Atheism, in Darwin's view, was all well and good for the intelligentsia, but ordinary people were not yet "ripe" for atheism. So he called himself an agnostic, largely for diplomatic reasons.'*

---

* Darwin's son Francis noted, in *The Life and Letters of Charles Darwin* (1887), that 'Dr. Aveling has published an account of a conversation with my father. I think that the readers of this pamphlet ('The Religious Views of Charles Darwin', Free Thought Publishing Company, 1883) may be misled into see-

In making this claim, Dawkins was responding to the tenor of Bunting's article in which she had quoted Mark Pallen, professor of microbial genomics at the University of Birmingham and author of *The Rough Guide to Evolution*: 'A defence of evolution doesn't have to get entangled in atheism.'* Madeleine Bunting's aim was to question the appropriation of the Darwin anniversary celebrations to support the 'new atheist' call to arms. She suggests that 'in particular, what would have baffled Darwin is his recruitment as a standard bearer for atheism in the 21st century.' This is clearly a position with which Dawkins disagrees. Evolutionary thought is key to his anti-religious arguments, and an atheist Darwin obviously fits more easily within that agenda. Not only is Dawkins uncomfortable with an agnostic Darwin, but he even disapproves of nonbelievers like me tolerating religion.

---

ing more resemblance than really existed between the positions of my father and Dr. Aveling: and I say this in spite of my conviction that Dr. Aveling gives quite fairly his impressions of my father's views. Dr. Aveling tried to show that the terms 'Agnostic' and 'Atheist' were practically equivalent – that an atheist is one who, without denying the existence of God, is without God, inasmuch as he is unconvinced of the existence of a Deity. My father's replies implied his preference for the unaggressive attitude of an Agnostic. Dr. Aveling seems (p. 5) to regard the absence of aggressiveness in my father's views as distinguishing them in an unessential manner from his own. But, in my judgment, it is precisely differences of this kind which distinguish him so completely from the class of thinkers to which Dr. Aveling belongs.'

* A quote at which Mark Pallen has revealed his amusement on his *Rough Guide to Evolution* blog. He was apparently arguing that we should concentrate on Darwin's positive achievements as a travel writer, geologist, zoologist, botanist and exponent of evolution, rather than become sidetracked. He also pointed out that atheism doesn't need Darwin anyway.

Butting out

In an article entitled 'I'm an atheist, BUT ...' on richarddawkins. net,* Dawkins has drawn up five caricatures of atheists who might choose to disagree with him:

1. 'I'm an atheist, but religion is here to stay.'
2. 'I'm an atheist, but people *need* religion.'
3. 'I'm an atheist, but religion is one of the glories of human culture.'
4. 'I'm an atheist, but you are only preaching to the choir. What's the point?'
5. 'I'm an atheist, but I wish to dissociate myself from your intemperately strong language.'

For Dawkins, No 1s are smug, self-satisfied pseudo-realists and No 3s are smirking fetishists who fancy the cultural riches of religion but are blind to its cruelties, while No 2s are merely patronising elitists who think it's OK for the proles to believe in religion, but it is not something they would indulge in themselves.

But it is the final two that I find the most interesting coming from someone who promotes himself as an ambassador of science.

Choir practice

His riposte to No 4s, those who question the purpose of challenging religion in such a polarised way, illustrates perfectly how Dawkins sees himself as a progressive liberator. He makes the analogy with feminists who 'preached to the choir' when they argued against the use of sexist pronouns. The point was, he said,

---

* Some of these points are also in the introductory preface to the paperback edition of *The God Delusion*.

to raise consciousness among those already aware of the issues by bringing non-sexist pronouns into everyday language.

So it is with atheism, Dawkins insists: 'Atheists as well as theists unconsciously buy into our society's convention that religion has uniquely privileged status.' And with a politician's adroitness, he links his argument on atheism to the moral outrage against sexist language: 'However right-on we may have been on the political issues of rights and discrimination, we nevertheless still unconsciously bought into linguistic conventions that made half the human race feel excluded.' But making a Freudian association between strident atheism and women's rights is no more valid than the activities of cigarette companies in the 1920s who hired models to smoke on suffragette marches, in order to implant in people's minds the idea that smoking was a badge of independent womanhood. And it is, of course, complete nonsense to imply even by association that atheists face the same battle against discrimination and exploitation that women do.

I would contend, too, that it is not even effective tactically. The concentration on 'sexist' language against women has sometimes distracted from the real and still unfortunately prevalent discrimination issues, and left us with Disgusted of Tunbridge Wells writing letters to the *Telegraph* guffawing at 'political correctness gone mad'. So Dawkins' 'consciousness-raising' campaign may not be the best way to promote the acceptance of rational science, or even atheism for that matter. Ultimately, it tends to raise hackles rather than consciousness.

Strong words

Which brings me to No 5s, the mealy-mouthed shrinking violets who do not appreciate his intemperate language. Quoting his old friend Douglas Adams, Dawkins argues that politicians, book reviewers and restaurant critics *et al* use intemperate language, so why shouldn't he? However, what is fine for a theatre critic trying to sell newspapers with his entertaining rants or a politician touting for your vote – or a fire-and-brimstone preacher trying to frighten the vulnerable into church – is surely not so acceptable for someone who promotes himself as a spokesman for science. I would say that science communication should aim for balanced debate that allows people to engage with the often complex ideas involved on their own terms. Using intemperate language or shock tactics in this context only serves to further entrench people's positions rather than encouraging them to become involved in open discussion. Dawkins himself takes issue with others who use shock tactics – as we saw in the previous chapter with his response to the recent 'Darwin Was Wrong' headline in *New Scientist*. The use of polemic, or even clever writing, has long been questioned by scientists. As far back as the seventeenth century, Francis Bacon argued that the power of their own rhetoric could seduce scientists away from a real understanding of nature:

> Men began to hunt more after words than matter – more after the choiceness of the phrase, and the round and clean composition of the sentence, and the sweet falling of the clauses, and the varying and illustration of their works with tropes and figures, than after the weight of matter, worth of subject, soundness of argument, life of invention, or depth of judgment.

Relative values

We shall return to the issue of how science should be communicated and Dawkins' belief in advocacy later. But it is worth answering his contention that the 'I-am-an-atheist-but' response to his work is guilty of 'cultural relativism' of the worst kind. For example, in his 'I'm an atheist, BUT …' article, Dawkins recalled an atheist critic who approached him when he was on a book tour in the US. This critic apparently fitted into the camp of those who see religion as a glory of human culture. In response to this Dawkins approvingly quoted psychologist Nicholas Humphrey's horrified reaction to a television documentary about the religious sacrifice of a young Inca girl: 'The message of the television programme was in effect that the practice of human sacrifice was in its own way a glorious cultural invention – another jewel in the crown of multiculturalism.'* He also used this example in chapter 9 of *The God Delusion*, and went on to suggest that 'decent liberal readers', meaning those who believe that we should judge the actions by Inca standards rather than our own, may feel uneasy with Humphrey's response. This is a slightly disingenuous example that does not effectively counter his critic.

No one I know, however 'liberal' they may be, would feel anything but revulsion at the prospect of the slaughter of a little girl today or historically, whether for religious reasons or any other, and it is patently ridiculous to imply that anyone who doesn't agree with everything that Dawkins says would. It is also absurd to

---

* Dawkins sums up his analysis of his encounter with this critic by saying: 'It would be unfair to accuse our critic in San Diego of complicity in such an odious attitude towards the Inca "ice maiden". But I hope at least he will think twice before repeating that *bon mot* (as he obviously thought of it): "I believe in people, and people believe in God." I could have overlooked the patronizing condescension of his remark, if only he hadn't sounded so smugly *satisfied* by this lamentable state of affairs.' A number of Dawkins' supporters have claimed that criticisms of him tend to be personal attacks rather than critiques that deal effectively with his arguments; yet in this case he made exactly the same kind of *ad hominem* attack on a critical member of the public – which I believe was highly inappropriate.

condemn a need to understand cross-cultural perspectives on the role of religion in society just because the Incas practised human sacrifice long ago. This is another example of Dawkins' arguments by extremes that are wholly inappropriate in a scientist. By arguing in such an overly emotive way, Dawkins fails to engage with the more complex issues of intercultural debates.

This is problematic on two levels. First, it takes a very heavy-handed, naïve approach to a raft of very intricate problems and second, the implication is that if you don't agree with him you are condoning abhorrent practices. The upshot is to entrench a dogmatic approach which is far removed from any informed rational or scientific debate. Criticism of 'liberals' is a recurrent theme in Dawkins' later works. Yet one has to seriously wonder: if he is arguing against cultural relativism in the sciences and ethics – an argument that supposes that science transcends culture – why then does he keep dragging his personal politics into supposedly rational or scientific debates? Of course we feel moral outrage at any cruelty practised anywhere. And of course we have to take a stand against barbarity. But we don't condone such cruelties simply by attempting to understand them, just as we don't condone genocide by reading a biography of (the atheist) Stalin. Indeed, I would say that if any of us have any claim to moral enlightenment, and if we believe that we see things more clearly and fairly than others do, it is only because of our willingness to try to understand and engage with other points of view rather than condemn out of hand. Dawkins' unsophisticated approach to cultural debates could be a dangerous attitude to take. In part it may come from his response to the terrorist attacks of 11 September 2001. These clearly had a profound influence on him, as he has mentioned them in a number of interviews and the preface to *The God Delusion*.

Just four days afterwards, on 15 September 2001, Dawkins wrote an article in the *Guardian* about the 9/11 attacks. While understandably shocked by the whole affair, he took no prisoners in his criticism of the role of religion in the attacks:

> I am trying to call attention to the elephant in the room that everybody is too polite – or too devout – to notice: religion, and specifically the devaluing effect that religion has on human life. I don't mean devaluing the life of others (though it can do that too), but devaluing one's own life. Religion teaches the dangerous nonsense that death is not the end.

He surmises that 'To fill a world with religion, or religions of the Abrahamic kind, is like littering the streets with loaded guns. Do not be surprised if they are used.' To reduce the intense geopolitical situation to a clash of cultures – us versus them, the modern rational West versus medieval Islam – is a gross misrepresentation at best, and at worst a highly offensive and counterproductive stance. I would genuinely like to know why an ethologist who at the time held a chair for the public understanding of science was seen as someone who should be making such grand political and subjective statements in public.

It is right that we should fight against fundamentalism in all its guises, but it is self-evidently impossible to fight fundamentalism and dogma with fundamentalism and dogma. What is needed is open debate and dialogue that enables mutual understanding, not rhetoric that only serves to entrench polarisation.

The problem with Dawkins' public persona is that he rarely reflects his criticisms of others back onto himself. This can lead to him making apparently conflicting statements. In a 2003

Guardian article about the Iraq invasion, in which he was highly critical of George W. Bush, Dawkins argued that the logic of the invasion was based on revenge for 9/11 and was 'pure racism and/ or religious prejudice'. How does he reconcile this with his own personal, often intemperate, attacks on religion? A classic example of this was when, in the preface to *The God Delusion*, Dawkins related how pleased he was with the Channel 4 advertisements for his television programme *The Root of All Evil?* 'It was', he tells us, 'a picture of the Manhattan skyline with the caption "Imagine a world without religion". What is the connection? The twin towers of the World Trade Center were conspicuously present.'

Cultural nuances

While expressing outrage about human sacrifice or even terrorist attacks is an effective way to get people on your side, it's intellectually (and morally) lazy to use this as an argument against trying to understand the morality of other cultures.

Neither is it a particularly sophisticated approach to denouncing cultural relativism. This is the school of thought that says we should judge the actions of individuals or societies within different cultures on their own terms and through the perspective of their own practices or customs. In essence, it is a viewpoint that suggests that there is no objective way to evaluate culture. A cultural relativist might argue that a moral stance that is unacceptable in one culture should be respected or carry equal weight if it has historically been taken, or is currently taken, in another culture. An extension of this is that a relativist might, for example, argue that 'Western science' has no privileged position in helping us to understand the world – it could and should have the same status as the practice of witchcraft in some tribal cultures. It is easy to see why Dawkins is not enamoured of such arguments, and while I understand the philosophical debates behind them, if I am honest I am not particularly convinced by them either. But rejecting the extreme cultural relativist position is far removed from recognising that we need to engage in cross-cultural dialogue.

An example of this 'cultural relativism gone mad' that Dawkins cited in *The God Delusion* is female genital mutilation (FGM). Following on from his comments about Inca priests, he suggested that there is a 'conflict in the minds of nice liberal people who, on the one hand, cannot bear suffering and cruelty, but on the other hand have been trained by postmodernists and relativists to respect other cultures no less than their own.' Here Dawkins suggested that female genital mutilation is regularly practised in the UK and that local authorities ignore the practice 'for fear of being thought racist "in the community"'. It should be noted, though, that he only provided anecdotal evidence for this, and that it is illegal for young women to be circumcised in the UK or taken abroad for the purpose of undergoing this procedure. So there is clearly no hand-wringing postmodernism evident in UK legislation on the matter.

Contrary to received opinion, female genital mutilation is not a 'religious' activity – it is a culturally and socially ingrained practice which has been undertaken in some regions for centuries. It is true that in certain areas it may appear to be linked to religious groupings. However, the real picture is far more complex and dependent on a number of factors within various communities. As two recent studies – published in 2004 and 2006 – show, there is no specific religion (of any ilk; be it local traditional religion, Islam or Christianity) that teaches that female genital mutilation is a necessary facet of religious doctrine. Certainly, neither the Bible nor the Qur'an state that female genital mutilation should be practised. Its prevalence is linked to wider socio-economic factors such as access to education. It may even be, in the longer term, that involving religious leaders in condemning the practice will be a factor in its eventual eradication. However, in typically disingenuous fashion Dawkins implies that FGM is simply an 'ethnic religious habit' that those fictitious woolly-minded liberals agonise over but see as quaint.

If we are to engage in coherent discussions about the ethical dimensions of cultural debate, then surely we need a slightly less ham-fisted approach than the one Dawkins has employed on this subject. He himself, as we will discuss shortly, clearly agrees that morality is the province of philosophy. That being the case, he should arguably not use the platform of being a scientist and science communicator to make what are clearly philosophical and to some extent political points.

A very public opinion

All this matters not because I do not think Dawkins is entitled to his point of view, but because it has a huge impact on the way in which he presents the case for evolution, and an equally huge impact on the public perception of science. In a recent Channel 4 documentary that Dawkins presented called *The Genius of Charles Darwin*, he referred to a survey that apparently showed that 40 per cent of people in the UK believed in creationism. Alarming evidence that Dawkins' forthrightness on evolution is needed to stem the tide of religion, it would seem.

Yet Dawkins at no point makes it clear which survey this was. So I've tried to track it down, and it seems to have been a poll undertaken for a programme in the BBC documentary series *Horizon*. It is worth noting that the title of the programme was *Horizon: A War on Science*, and it's fairly apparent that the programme-makers approached their poll with certain expectations. They were clearly not conducting a scientific study but merely attempting to gauge public opinion and attract interest in their upcoming programme. There is nothing at all wrong with this, but a scientist would be wise to be a little more cautious in uncritically quoting their findings.

Polls apart

The poll was conducted between 5 and 10 January 2006; the programme was transmitted on 26 January 2006. It is fairly safe to assume that the programme had already been virtually, if not entirely, finished before the survey was conducted, and so the purpose of the survey was almost certainly to generate publicity. The BBC press release states:

> Participants in the survey were read three statements and asked which best described their view of the origin and development of life. The statements were:
>
> - the 'evolution theory' says that humankind has developed over millions of years from less advanced forms of life. God had no part in this process;
> - the 'creationism theory' says that God created humankind pretty much in his/her present form at one time within the last 10,000 years;
> - and the 'intelligent design' theory says that certain features of living things are best explained by the intervention of a supernatural being, e.g. God.*

---

* http://www.bbc.co.uk/pressoffice/pressreleases/stories/2006/01_january/26/horizon.shtml

A point I'd like to pick up on here is the glossing over of the difference between the 'origin of species' and the 'origin of life'. The origin of species, of course, is about how species have emerged over the long expanse of geological time; the origin of life (or abiogenesis) on the other hand is how life started in the first place. These are two different theories and areas of research. Although Dawkins does weave origin-of-life discussions into his proto-selective process, particularly in the book *The Ancestors' Tale* which was written in conjunction with his able research assistant Yan Wong, the origin of species is essentially the province of Darwinian evolutionary theory while the origin of life is not. That is not to say that there are no cogent scientific theories about the origin of life, but when it comes to gauging public opinion about evolutionary theory this distinction can make all the difference.

Many people of faith happily hold the beliefs both that life evolved entirely by natural selection and that God started it – in other words, a theistic hands-on origin of life and an evolutionary hands-off origin of species. God as the 'first cause', if you will. But the questions outlined in the poll do not allow for this view. The choices are very much either/or, provide no room for subtlety and clearly promote the idea that there is an inherent antagonism between evolution and religion. A person of any faith would find it extremely difficult to choose the first option, as it states that God had no part in the origin of life. Yet the other two options force them to either choose creationism or intelligent design, or simply say 'I don't know'.

Lesson plans

So what exactly did the survey show? The results are interesting. As the BBC reported:

Of those surveyed, 48% said evolution theory most closely describes their view; 22% chose creationism; and 17% chose intelligent design.

A further 12% said they did not know. This brings the total percentage of those not choosing evolution to 51%.

When asked if the theories should be taught in school science classes, 69% agreed that evolution should be taught; 44% that creationism should be taught; and 41% that they believed intelligent design should be included on the science curriculum.

The headline of the BBC news website's analysis of this survey starts with the apparently worrying statement for scientists that:

> Just under half of Britons accept the theory of evolution as the best description for the development of life, according to an opinion poll. Furthermore, more than 40% of those questioned believe that creationism or intelligent design (ID) should be taught in school science lessons.

So less than half of Britons support evolution and almost half believe that the fiction of creationism and intelligent design should be taught in *science* lessons, even though they are not scientific. If we took this entirely at face value, it would suggest that scientists do indeed face a major struggle against a tide of religious prejudice – all the more worrying in that it would represent a step back from half a century ago.

Christian scientists?

Yet this survey may not be as clear-cut as it seems. If only 48 per cent of the people surveyed accepted the theory of evolution, for instance, then why did 69 per cent of those polled think it should be taught in schools? Even allowing for all the 'don't knows' to be added to those who chose to accept evolutionary theory – that's still only 60 per cent of those surveyed. And if, as is suggested, 51 per cent of those surveyed rejected evolutionary theory, why would such a significant proportion of these people choose to have evolution taught in school science lessons?

Moreover the *Horizon* editor went on to make the common mistake we saw before of conflating 'evolutionary theory' with the 'origin of life', by commenting on the survey that: 'Most people would have expected the public to go for evolution theory, but it seems there are lots of people who appear to believe in an alternative theory for life's origins.' I am not criticising the BBC here, who rightly said of the poll: 'I think that this poll represents our first introduction to the British public's views on this issue.' Their intention was simply to arouse interest, but it does concern me that – if my suspicions are right about which poll Dawkins was referring to – he and others might use such a survey to promulgate an atheist call to arms.

Regardless of the way in which the results of this survey are interpreted, it does still suggest that a surprising number of people seem to think that evolutionary science should not be taught in schools. However, a much more refined longitudinal study is clearly needed. I would hazard an educated guess that if such a survey were repeated without leading questions, the results would paint a much less stark and dichotomous picture. My point is that this little foray into the world of polls and statistics has illustrated the way in which theistic or deistic acceptance of evolution and 'creationism' can easily be conflated in the popular conception of the debates surrounding science and religion.

Classroom assault

The controversies over evolution and education are always headline-grabbers, and the temperature of coverage of the subjects in the media has certainly not encouraged a balanced approach. Moreover, well-funded religious groups in both the USA and UK have begun to use digital media and modern publicity techniques to spread their anti-evolution arguments into schools in a quite disturbing way. In the UK, groups like Truth in Science have in previous years sent out 'teaching packs' designed to be used to teach the 'alternatives' to Darwinism, not as part of religious education but as part of the *science* curriculum. The classic example of this, of course, is the 'intelligent design' line, which frames creationism as a 'scientific' theory, using 'scientific'-style arguments, but simply missing out the crucial element of science – that of studying *all* the evidence objectively. This undermines not only science but also theology.

This is all very worrying, and scientists need to work hard to prevent this retrograde blurring of science and religion. However, I find it equally troubling that in fighting against it, Dawkins seems to be adopting the same dubious tactics. On the Richard Dawkins Foundation for Reason and Science website, under the heading 'Our Mission', you will find the following evangelical-sounding statements of intention:

Research: We intend to sponsor research into the psychological basis of unreason. What is it about human psychology that predisposes people to find astrology more appealing than astronomy? At what age are young people most vulnerable to unreason? What are the correlations between religiosity and superstition on the one hand, and intelligence, educational level, type of education etc. on the other? Research of this kind would be supported in the form of grants to universities in America and Britain or wherever the best research can be done.

Education: Within the limits imposed by the charity laws of the respective countries, we would seek to support rational and scientific education at all ages, and to oppose the subversion of scientific education, for example by the well-financed efforts to teach creationism in science classes. Depending on how much money we raise, we would hope to subsidize the publication of books, pamphlets, DVDs and other educational materials.

Are we to fight fears of indoctrination on both sides of this debate by funding research into the age at which children are most susceptible to indoctrination, or asking 'at what age are young people most vulnerable to unreason'? It raises the question: who defines 'unreason'? If we were to read the converse on an evangelical religious website we would rightly question it. It is of course perfectly acceptable to create educational materials on evolutionary thought, but framing these as a way of 'opposing the subversion of scientific education' would only be counter-productive, alienating many people and reinforcing the supposed clash between science and religion. Would this not just be the same kind of proselytising that he seeks to oppose?

Some pressure groups argue that we should teach the 'controversy' about evolution, intelligent design and creationism in the science classroom – a view which I myself, and many scientists, find problematic. Yet in the same television series in which Dawkins quoted the survey about public attitudes to evolution, *The Genius of Charles Darwin,* there is a jaw-dropping episode in which he goes into a classroom to discuss evolution. Does he simply teach the facts? No, he frames the entire debate about Darwin and evolution in terms of a conflict between science and religion – he teaches the 'controversy' – exactly what he is purportedly arguing against. Surely, Dawkins would agree that the science classroom is not the place for debates about religion – so why attempt to teach Darwinism by framing it in terms of religion?

Denying religion

If we are to believe Dawkins, on the one side there is moral certainty and the clear-sighted logic of science and atheism, evolution and the selfish gene, inseparable and incontrovertible; and on the other moral ambiguity and the delusions of religion and creationism. There are, for him, no other possibilities. He rejects entirely, for instance, the arguments of Stephen Jay Gould in his book *Rock of Ages* that science and religion are 'non-overlapping magisteria' or NOMA. 'Science and religion are not in conflict, for their teachings occupy distinctly different domains', wrote Gould; '… I believe, with all my heart, in a respectful, even loving concordat'. Each magisterium, for Gould, is a domain where different types of knowledge are employed to resolve debates or conflicts. 'The net of science covers the empirical universe: what is it made of (fact) and why does it work this way (theory). The net of religion extends over questions of moral meaning and value.'

There are several reasons why Dawkins rejects this view. First, he argues that religion has no special claim to moral expertise, and I'd agree. Indeed, morality has never been religion's sole province and philosophers have always played a profound part in shaping our ideas about morality – though clearly there has been an interplay between the two throughout history. A large amount of what we might tend to think of as 'Judeo-Christian' morality, for instance, tends to spring from Greek philosophical thought, and early Christian writers were heavily influenced by the ethical ideas of, among others, Aristotle and Plato. And it cannot be denied that there is an uncanny similarity between Kant's categorical imperative and the Christian 'Do unto others as you would have others do unto you'.

In *The God Delusion*, Dawkins accepts that 'the question of "What is right and what is wrong?" is a genuinely difficult question that science certainly cannot answer'. He elsewhere argues: 'Unfortunately, the hope that religion might provide a bedrock, from which our otherwise sand-based morals can be derived, is a forlorn one.'* On these points he argues, and I concur, that our approach to morality should be based in sound philosophical reasoning and does not fall within the realms of either 'science' or 'religion'. Clearly an understanding of both science and culture plays an important part; however, both of these have only a partial role when it comes to making complex ethical judgments or understanding the basis of morality.

Is religion scientific?

So far I'm with Dawkins, but then comes his principal reason for rejecting Gould's 'non-overlapping magisteria' thesis: because religion makes 'existence claims'. Dawkins argues that 'the Virgin Birth, the bodily Assumption of the Blessed Virgin Mary, the Resurrection of Jesus, the survival of our own souls after death: these are all claims of a clearly scientific nature'. He means that some people claim that these things really happen or happened, and so they fall within the province of science. 'Either Jesus had

---

* R. Dawkins, 'When Religion Steps on Science's Turf: The Alleged Separation Between the Two Is Not So Tidy', *Free Inquiry*, vol.18, no 2.

a corporeal father or he didn't', Dawkins says. 'This is not a question of "values" or "morals"; it is a question of sober fact. We may not have the evidence to answer it, but it is a scientific question, nevertheless. You may be sure that, if any evidence supporting the claim were discovered, the Vatican would not be reticent in promoting it.'

All this is true, and we are on relatively safe ground scientifically in saying that human virgins don't give birth and that ordinarily dead people don't come back to life – because we have observed the contrary on many occasions.

However, the issue of whether or not souls survive after death is unlikely to be answered by science – we cannot know this empirically. While religion may make claims about the existence of souls, what may happen to them after death and so on – these are not necessarily trespasses into the realm of science. To assert that there is no such thing as a soul is just as much a statement of belief as to assert that they do exist – we cannot devise any experimental basis on which to make assertions as to what happens to 'consciousness' after death.

Dawkins' confusion

Dawkins seems to confuse what he can know empirically, or even to an extent rationally, with what he believes. He *knows* based on observational evidence that evolution occurs. He *knows* based on observational evidence that virgin women don't give birth and dead people don't come back to life. However, he does not *know* based on observational evidence that there is no soul or deity. He surmises based on his personal experience that there is no God – this is a different kind of proposition and open to much debate.

Now I am also an atheist* and a firm supporter of evolutionary science, and yet I can differentiate between these two things. The first – atheism – is my own personal conviction; the second – my understanding of evolution – is based on the abundant scientific evidence supporting the theory. We will discuss this distinction further in the next chapter. As enticing as it might be to claim science for atheism, this stance could ultimately undermine 'science' in the public perception by reaching beyond what are commonly accepted to be its parameters.

The common, and critical, mistake here is to unthinkingly accept the idea that evolutionary theory automatically disproves or discounts the existence of God. Dawkins is not an atheist merely because his acceptance of evolutionary theory will not allow otherwise. Evolutionary theory, and indeed science, makes no claims either way about the existence of a deity; even though it supports the theory that natural laws are responsible for the development of new species over a vast period of geological time. Dawkins is an atheist because he chooses to believe there is no deity. He even, laudably, admits himself that he cannot say with certainty that there is no God, only that there is probably no God. Unfortunately, I think that this important point sometimes gets lost in the din of his otherwise black-and-white portrayal of the relationship between science and religion. I am not entirely sure how one would devise a scientific experiment to disprove 'God' – but I bet it would need a rather large budget!

In a recent book on Dawkins and religion, *Dawkins' God*, Oxford theology professor Alister McGrath argues that 'an

---

* Though this is a term I hesitate to use thanks to Dawkins' appropriation of it – the key point being that I don't see my own non-belief as being in opposition to anyone else's belief.

evidence-based approach to [the relation of both science and religion] is much more complex and much more interesting than Dawkins' "path of simplicity and straight-thinking.'"* And while some of Dawkins' arguments do seem compelling, they do rather ride roughshod over the more complex historical and theological aspects of the debate.

The idea that there is not necessarily a conflict between science and religion is by no means a new one. Even after the much romanticised and now infamous clash between Galileo and the Catholic Church which has always been billed, misleadingly, as a simple conflict between science and religion, there was a continued reconciliation between 'science' and 'religion'. The general rule of thumb used by the Catholic Church around the time of Galileo was to adopt a sort of instrumentalist approach. Theories about how the universe worked were fine in that they were effective ways of predicting phenomena, but this did not make them 'true' realist representations of the universe. Thus the Catholic Church was generally able to accept scientific ideas, even when they seemed to conflict with Biblical doctrine, by accepting them as valid ways of looking at the universe without necessarily being 'true' in the ultimate metaphysical sense, and to some extent seeing the Bible as allegorical rather than literal truth. Meanwhile, as we saw earlier, deists and Anglican natural theologians were able to come to terms with a scientific view of nature which had God 'behind the scenes'.

---

* McGrath maintains that Dawkins is 'ignorant' of Christian theology, and so cannot deal with questions of religion and faith intelligently. In reply, Dawkins asks: 'Do you have to read up on leprechology before disbelieving in leprechauns?' It's an amusing if polemical response, but hardly the words of a scientist approaching what he himself refers to as religion's scientific claims.

No place for accommodationists

Dawkins simply does not have time for those who manage to both hold religious views and accept evolutionary science; and neither does he have much patience with 'accommodationists', atheists like me who have no wish to pick a fight unnecessarily. In his typically belligerent fashion, Dawkins describes the situation thus: 'Scientists divide into two camps over this issue: the accommodationists, who "respect" creationists while disagreeing with them; and the rest of us, who see no reason to respect ignorance or stupidity.' Dawkins simply can't resist being offensive and dogmatically right. I happen to disagree profoundly with creationists' incursions into the realm of science, but I also think that it really doesn't help the cause of science to call them stupid. It is also worth noting that accommodating people's religious beliefs does not necessarily mean accommodating creationism.

This remark about creationists being ignorant and stupid came in response to the resignation of Professor Michael Reiss as Director of Education of the Royal Society, an appointment which, when first made, since Reiss is an ordained minister Dawkins described thus: 'A clergyman in charge of education for the country's leading scientific organisation – it's a Monty Python sketch.' Professor Reiss resigned after reportedly saying that creationism should be taught in classrooms in a presentation at the British Association Festival of Science in September 2008. In fact this is not quite what he said,* but it gave ammunition to

---

* I was at that session and heard the paper he gave. I certainly didn't come away with the impression that he was advocating the teaching of creationism, or even saying that it should be brought up during science lessons by teachers. He merely, to my mind and that of others there, was arguing that one should not deride children who bring up their own creationist beliefs in lessons. Far from supporting the view that creationism should be given equal teaching time or even that we should 'teach the controversy', I interpreted it more as an argu-

those who accused him of being an 'accommodationist', and he had little choice but to go. Dawkins' spleen was not just vented on accommodationists, however. His double-barrelled shotgun opened fire on much bigger targets – creationists themselves:

> Get the bishops and theologians on the side of science – so the argument runs – and they'll be valuable allies against the naïve creationists (who probably include the majority of Christians and certainly almost all Muslims, by the way).*

Apart from the fact that it seems both bad for 'science's' reputation and unnecessarily offensive to infer that most Christians and almost all Muslims are 'ignorant' and 'stupid', it is not even necessarily true that they are 'naïve creationists'. It is shocking enough that Dawkins forgets science's massive debt to both Islamic and Christian scientists in the past and present. But it is just as irresponsible for someone who claims to represent science to make such a blatantly unscientific inference.

Naïve creationists?

What evidence does Dawkins employ when he asserts that the majority of Christians are creationists – let alone 'naïve' creationists? In fact, a large recent survey conducted by the theological thinktank Theos in February 2009 showed that, of the Christian respondents, 37 per cent responded positively to a 'theistic' version

---

ment that it is counter-productive to behave in a derogatory manner in the classroom towards children's beliefs.

* R. Dawkins, 2008. 'Letter to New Scientist on Royal Society Row', *New Scientist*, 18 September 2008. http://www.newscientist.com/channel/opinion/dn14748-letter-richard-dawkins-on-the-royal-society-row.html

of evolution, 37 per cent responded to a 'young earth creationism' option and – most interestingly of all – 5 per cent responded positively to a 'atheistic evolution' option.* And there is as yet little concrete research evidence to suggest that 'almost all Muslims' are naïve creationists as Dawkins asserts.[†]

Some of the initial research that has been done suggests that the more political 'Islamic creationism' is a cultural import from the US. The most recent and well-publicised incursion of 'Islamic creationism' was the distribution of the *Atlas of Creation* by Turkish creationist Harun Yahya. A huge, lavishly illustrated but completely unscientific 'text' book, the *Atlas* was sent out to thousands of academics in 2006. It used a familiar tactic of misquoting from scientists or science articles in an attempt to support its outlandish claims, which are akin to intelligent design. Harun Yahya's book has no doubt raised the profile of 'Islamic creationism', and may be one of the reasons for Dawkins' comment that 'certainly almost all Muslims' are 'naïve creationists'.

There has actually been very little research published to date in this area; however, one study – 'Public Acceptance of Evolution', published in *Science* in August 2006 – does suggest that of the countries surveyed, 'only Turkish adults were less likely to accept the concept of evolution than American adults'. However, there

---

* The possible responses were: young earth creationist option (humans were created by God sometime within the last 10,000 years); intelligent design option (humans evolved by a process of evolution which required the special intervention of God or a higher power at key stages); theistic evolution option (humans evolved by a process of evolution which can be seen as part of God's plan); and atheistic evolution option (humans evolved by a process of evolution which removes any need for God).
[†] In the Theos poll, for example, the number of Muslims (124 in a sample size of 2,000) polled was so small that they are only indicative rather than conclusive.

are a multitude of reasons why this might be the case, some of which have very little to do with a simple clash between 'science' and 'religion', and some of which are very heavily linked to both the historical promotion of 'Darwinism' as 'atheism' in the nineteenth and twentieth centuries, and access to resources on the subject of evolution. We should not be tempted to reduce the entire historical, philosophical, socioeconomic, educational and geopolitical debate to a bald statement that 'certainly almost all Muslims' are 'naïve creationists'.

A more recent paper published in December 2008 in *Science* points out that 'just as there is no monolithic Islam, there is no "official" opinion on evolution'. While highlighting the very real issues relating to poor public understanding of the science and the relatively low levels of acceptance of evolutionary theory in Kazakhstan, Turkey, Indonesia, Pakistan, Malaysia and Egypt, the paper's author Salman Hameed encouragingly says that 'although the survey results may point to a dire situation, the reality on the ground is more complicated. Evolutionary biology is included in the high school curricula of many Muslim countries.' However, he also argues that any attempts at communicating evolution today 'that link evolution with atheism will cut short the dialogue, and a vast majority of Muslims will reject evolution'.

Again, though, research into the comparative acceptance of evolutionary theory over time and across cultures is an area that needs much more research, and in the meantime it is probably advisable to refrain from making grand assertions as Dawkins does. We all need to be careful that we do not allow narrow-mindedness into this debate. It would be folly to assume that one individual, the creationist Harun Yahya, is representative of the majority of Muslims, in the same way that it is wrong to assume that Becky Fischer (the director of the now discontinued and highly controversial Jesus Camp in the US) is representative of Christianity. And there are several atheists and scientists who do not think that Dawkins represents their viewpoint.

Scientists play God

Why should someone who is a self-avowed atheist stick up for those of faith? Well, there is much more at stake than a simple turf war between science and religion, especially given the current threats to both the environment and biodiversity. It is only through reliable, trustworthy and adequate communication of often complex scientific ideas that these issues can truly be tackled on a global scale. I genuinely think it is important to leave old prejudices aside when facing a problem of such immense proportions that it will affect us all. A problem to which the application and understanding of 'science' may very well hold the key.

There is no doubt that both science in general and Darwinism in particular can provoke fear and distrust. Ever since Mary Shelley's *Frankenstein*, many have felt that scientists tend to overreach themselves. The image of the ambitious scientist has long been linked with hubris in the public perception. Unfortunately, this seemed to come true in the very worst possible way in the eugenicists' misinterpretation of 'evolution'. Just as we should not forget how the Jews and other races were persecuted by the Nazis, so we should not forget that science and the misuse of science has bred its own horrible follies. Mistakes such as the thalidomide tragedy have further reminded the public that scientific advances may not always be a straightforward blessing. So before we supporters of science accuse others of ignorance and stupidity, we should be absolutely sure that we are not also blameworthy in this respect.

The public is particularly wary of science being used by governments and other organisations in a political or otherwise dubious way, with conspiracy theories and distrust of advances such as genetic modification rife. There was no more clear-cut example of this growing wariness than the anti-vaccination movement's response to the measles, mumps and rubella (MMR) vaccine. As we will explore in the next chapter, scandals such as these that result in the public losing trust in science can have tragic consequences. Against this backdrop, to confuse and obfuscate what we can and cannot say with science is highly problematic.

It is for reasons such as these that we need to keep a level head and not fire off at our 'opponents' without thinking. The consequences are potentially just too severe. The aim of science communication surely should be to equip people with the critical and analytical skills to educate themselves about science, *not* to dictate what they should believe. Since the publication of *The God Delusion* Dawkins has often been charged with being an unwitting recruiter for creationism. A damning indictment that must surely have hit home. Tellingly, in a recent interview with the BBC's Owen Bennett-Jones, Dawkins seemed to be softening his line:

> Many people do [believe Darwinism is compatible with belief in God], because there are plenty of clergymen, bishops, theologians and things who of course go along with evolution. They have no choice; the evidence is overwhelming. I personally think it's rather difficult, but that's my personal opinion and you'll find plenty of clergymen to disagree.

He even went on to acknowledge that Darwin only ever referred to himself as an agnostic, rather than an atheist. I sincerely hope he continues in this more measured vein.

CHAPTER 7

SCIENCE MATTERS

A little learning is a dangerous thing. This has never struck me as a particularly profound or wise remark, but it comes into its own in the special case where the little learning is in philosophy.
Richard Dawkins, *A Devil's Chaplain*

One of the more commonly repeated complaints about Dawkins in recent years is that he has become more and more like those he criticises – a rigid and dogmatic fundamentalist venting his atheist ire at every available opportunity. He is also often accused of picking on easy targets – those who represent the badly thought out extreme perspectives of faith or, as in his 2007 television programme *The Enemies of Reason*, those who are unthinking supporters of wishy washy new age mysticism. The charge is that he has focused only on people who couldn't string a cogent argument together if their life depended on it.

I think that this is perhaps a little unfair to both Dawkins and those he has famously interviewed. It is likely that his television series are an inevitable casualty of the cutting-room floor. I know from one interviewee on one of Dawkins' TV series that he was questioned on camera for over an hour and the end result was a far from flattering two minutes. It is also unhelpful to bandy around words like 'fundamentalist'. While he may be dogmatic in his approach to certain aspects of evolutionary biology and science communication, he is certainly not a fundamentalist. As we saw at the end of the last chapter, he is clearly capable of changing his arguments in light of certain criticisms – this is not the action of a fundamentalist.

In order to analyse Dawkins' work we need a clearer understanding not only of the history of evolutionary science and current developments in the field, as we have explored in earlier chapters, but also of what we mean when we talk about science. Dawkins' 'dogmatism' springs not only from his confrontation of 'fundamentalist' positions he opposes, but also from his representation of the history and philosophy of science from within a narrow perspective on current debates in evolutionary biology. It is clearly important that as a society we question unfounded assertions or beliefs – especially when they concern issues that can have a detrimental impact on society, such as human health. However, Dawkins' forays into philosophy are one of the things that seem to attract particular public criticism. It has been repeatedly said online – even on the forums of Richard Dawkins' own website – that one of his main failings is that he has a poor grasp not of theology but of the philosophy of science. By providing a myopic and personal perspective on what 'science' is, Dawkins can sometimes undermine some of the genuinely good work he has done.

Science as a worldview?

In the preface to *The Blind Watchmaker* Dawkins says that the book is not a 'dispassionate scientific treatise' but explains that he wants 'to persuade the reader, not just that the Darwinian worldview happens to be true, but that it is the only known theory that could, in principle, solve the mystery of our existence.' While I don't doubt that evolutionary theory is the best explanation, Dawkins is actually describing *his* version of a neo-Darwinian worldview. As we will see in chapter 9 this is an interesting, and not altogether positive, approach to science communication. But it is Dawkins' version of what 'true' represents that we shall explore here.

Dawkins is sometimes described as being philosophical in his approach to the science/religion debate, and he himself claims his representation of science is that of an enlightened rationalist. In *The God Delusion* he states: 'God's existence or non-existence is a scientific fact about the universe, discoverable in principle if not in practice.' Dawkins clearly has a notion of what we are referring to when we talk of scientific facts. It is worth exploring what he really means here and why it matters. What we call a 'scientific fact' is a subject of much debate. This is clearly very different from the way in which we generally use the word 'fact' in day-to-day life. The term is tricky to define. We might, for example, say that 'it is a fact that Sarah is tired' but that is contingent – it is a description of a state of affairs. It describes the *circumstance* of Sarah being tired. Another way that we might use 'fact' is to describe things which are not contingent, i.e. that $2 + 2 = 4$. We tend to think of a 'fact' as a statement which goes hand in hand with the concept of truth. There are also different ways of using the words 'fact' and 'truth' depending on whether the matter is being approached from a scientific or a philosophical perspective.

In debates about how science works, this really comes down to what we call epistemology – or theories of knowledge. As we briefly discussed in chapter 1 there is a slight difference between how we might approach these ideas from a rationalist or an empirical/scientific perspective. And while these two viewpoints are not mutually exclusive, unpicking them can give us a helpful insight into the basis of some of Dawkins' claims for science. Dawkins is often described as a 'rationalist', but this is used in a different sense to the way rationalists in the history of philosophy might have used the term. With Dawkins, the terms 'rational', 'rationalist' or 'rationalism' merely mean 'uses reason', whereas rationalists in philosophy take a different approach to what 'rationalism' means and have been committed to a great deal more, including innate or *a priori* ideas which are not based on experience. This is distinct from the use of reason which is something common to most intellectual endeavours, especially science. Therefore this chapter will explore further what the term 'rationalism' means, and how it relates to the concepts of empiricism that underpin 'science' as we know it today.

What is science?

From a historical perspective it is very difficult to define the activity of 'science'. A lot of what we might understand as being the great scientific achievements of the past would not necessarily fit in with how we might try to distinguish science from other forms of knowledge today. There is plenty of debate about whether Darwin's work was 'scientific', or whether Darwin's theories are 'falsifiable' – a concept which we shall explore further. One thing you have to let go of quite quickly when studying the history or philosophy of science is an easily accessible, hard and fast definition of 'science', or of who we might count as a 'scientist'. An example is the way in which science works today: everything from funding to the way we define the term and the disciplines it encompasses, right through to the very institutions in which it is undertaken, is vastly different from how it was in Darwin's day. Even if we were to put the historical context of science to one side, it is still not easy to come up with an adequate description of what 'science' is. In addition, those who practise it can sometimes have very different perspectives on how it should be undertaken – this can be for a number of reasons such as disciplinary differences, technological capabilities or institutional traditions. For instance, should all science be based purely on experimentation? Some sciences use only observations, or are highly abstract and mathematical. Does science have to use mathematics? Again, some areas don't rely on the use of mathematics. 'Science' is a term we might use on a daily basis, but we should ask ourselves what it actually represents in terms of research activity and methodology.

To know or not to know, that is the question …

We could try to answer what science is by asking a series of linked questions:

1. What is it to 'know' something?
2. What then is knowledge?
3. What methods does science use to understand or gain knowledge about the natural world, and how are these different from those that we use every day to make sense of the world we live in?

If we start with the first question: there are a number of ways in which we might use the word *know*. In everyday language we tend to use the word when we feel that we are sure of something. You might, for example, say that you 'know' in the course of a dispute with someone, e.g. 'I know that Queen Victoria had an affair with her servant John Brown …' But this kind of argument may well be based on a mere second-hand opinion or unreliable information – here from the film *Mrs Brown*. In this case, as with many others, we are using the word 'know' to express a personal conviction, when our claim may in fact be groundless. We have been convinced by one perspective without having based our assertion on any evidence which supports the claim or proposition. Our source may be untrustworthy. Of course, trusting an authority or expert is not in itself a bad thing – in fact it is essential in the business of science, as not everyone knows everything. The mistake would be to pick someone who is not in fact an authority or expert in this case. It is perfectly acceptable if we are just having a discussion with our friends, but we don't really want scientists to base their ideas on such insufficient evidence. Given that knowledge implies some connection to truth, and that scientists cannot base their knowledge on such uncertain grounds, they can instead *claim* to have knowledge on these grounds.

Knowledge is not necessarily opposed to belief, but is instead belief that is justified and true. Obviously, in the sciences it is quite normal to suggest a hypothesis based on preliminary research – but this is usually a driver for new research rather than being seen as a definitive answer.

A classic example of the implications for science communication that arise from both these ways of using the word *know* is the measles, mumps and rubella (MMR) vaccine controversy. The scientific press statements that led to the recent MMR scandal in the UK were a speculative conclusion drawn from the results of a preliminary study of a very small sample size of twelve children. The scientists concerned did not necessarily have sufficient evidence to say that there was a link between the vaccine and the onset of autism, and their assertions were based on anecdotal reports from some of the parents involved in the research. Subsequent, more comprehensive and larger-scale research projects have so far shown no evidence of a link. One of the scientists proposed that there might be one, but his conclusion was speculative and really more research evidence was needed. He *believed* there was a link; however, this turned out to be an unjustified belief, and clearly we should not base our model of science on this kind of knowledge.

This kind of confusion about what we can and cannot claim to *know* in the realm of science can have huge implications for society. Understandably, this has contributed to a lack of trust in scientists. When there are conflicting accounts, it is difficult to know who to believe if you do not have access to the original research or the skills to interpret the data for yourself. Consequently, some people have chosen not to trust the view of the mainstream consensus in the scientific community that there is no recognised causal link between the MMR vaccine and autism. One unfortunate result of this is that in the UK today the uptake of the combined childhood MMR vaccine has fallen below the level where there is 'herd immunity' – as a consequence there have been repeated outbreaks of measles which have resulted in a number of children's deaths.

It is for this reason that scientists must be very careful about what they claim they can *know* scientifically. When scientists like Dawkins make claims that go beyond what we can really claim to know through scientific research – for example, that 'God's existence or non-existence is a scientific fact about the universe, discoverable in principle if not in practice' – it can unintentionally contribute to public confusion and distrust about what science can and cannot say. As we will see below, Dawkins' version of a scientific fact does not necessarily fit with some of the models we have to describe how we *know* things in science, and may indeed go beyond what some scientists see as being within the bounds of science.

What is knowledge?

There have been debates about how we know what we know about the natural world for centuries. The key question is how we might arrive at conclusions we can rely on. There are two key approaches that Dawkins uses – empiricism and a kind of neo-rationalism or scientific reasoning. In some ways he does so interchangeably, but while they are linked they describe very different approaches to the basis of knowledge.

A rationalist philosopher might mean something very different to a scientist when they discuss what it is to 'know' something. Philosophers tend to be concerned with the foundation of knowledge and its validity. When philosophers have approached the question of what it is to *know* something they have, in the main, been concerned with whether or not we can really 'know' anything, in the sense of possessing any information that is not open to question or doubt – as was outlined with Descartes in chapter 1. This is very different from the types of knowledge we have discussed above – in essence, justified beliefs *which may be wrong*. The rationalist philosopher might seek knowledge that we can know for *absolute certain* – not merely beyond reasonable doubt, but beyond any doubt. This has given rise to a set of problems that has occupied philosophers for centuries – because if you think about it there is very little, if anything, that we can know for *absolute certain*, beyond any doubt at all. And if we accept that we can know even a small amount for certain – how do we *know* what we know? Where does that leave the sciences – surely, if we can't know anything for certain, we might as well pack up and go home? The answer is no – as we shall see, science does not have to be based on this philosophical concept of certain knowledge – but the upshot is that scientists should not make claims to this kind of knowledge.

This brings us to one of the key ways in which Dawkins sometimes misrepresents what science can actually tell us about the world. He claims to be both a scientist and a rationalist, using scientific reason to argue his case. However, the boundaries between the two are often so blurred that it is sometimes easy for the reader of Dawkins' work to confuse 'science' and 'rationalism'. Of course it is perfectly feasible to be both a scientist and a rationalist. The problem arises, as in the science and religion debate, when the wrong set of tools is used to complete a specific job. I happen to personally think that there are some very compelling rationalist arguments for the probable non-existence of God, but as these are not based on scientific reasoning this does not make the probable non-existence of God a scientific fact as Dawkins claims.

This might seem to be a minor point but it is worth exploring – it is an important one to bear in mind when we consider how Dawkins represents his perspectives on evolutionary science, as we will see in the next two chapters. He has consistently blurred the boundaries between the physical and the metaphysical, and has subsequently been criticised for straying indiscriminately into the realms of philosophy. At this point we should briefly explore two schools of thought in the history of philosophy, namely rationalism and empiricism, in order to allow us to engage with the age-old philosophical question: what is knowledge?

Beyond reason

Science, as it is broadly conceived today, is a form of empiricism. Empiricism is the term used to describe those kinds of activity that attempt to construct an account of knowledge in terms of sense experience – it is based on observations of the natural world. In short, it is the application of observation and experiment. This is in some ways opposed to rationalism – which is the philosophical position that says that reason or logical thinking has precedence over all other ways of acquiring knowledge or, in its extreme form, that it is the only path to knowledge.

The key difference between the extremes of these two positions is that, in very simple terms, a rationalist would argue that we cannot be certain of our senses – of sight, hearing, touch or taste – and therefore neither can we be certain of the data which we observe through them. For example, in a rowing boat when the oars enter the water they appear bent, though that is clearly not the case; the optical effect is due to refraction of the light. But if our senses can deceive us in these circumstances, how can we be certain that we are not deceived at all times for reasons of which we are not aware? Historically, rationalists have argued that the only way we can discover absolute or certain *truths* about the universe is not by observing the world around us, but through the use of reason alone.

An empiricist, on the other hand, might argue that we can base knowledge on our sense experience, as we do not need 'certain' knowledge in order to understand – or to predict what might happen in – the world around us. The seventeenth century tends to be characterised by a number of great scientific breakthroughs (for example the work of Boyle, Hooke, Newton, Linnaeus and Buffon) through which people began to further realise their potential to control and use the physical world around them. The age-old quest of the philosophers for absolute knowledge became less important and some thinkers of the period said that they aimed not to discover real or certain truths about the universe, but only to develop probable hypotheses about the world. In this mould, science attempts to systematically collate the observable uniformities or regularities of the world around us in order to offer explanations and make predictions. For example, if you were to close this book now and let it drop out of your hand, I could fairly reliably predict that it would fall to the ground. (Unless of course you are in space, in which case you are cheating!)

People often think of science as a progressive attempt to collect these observable 'facts' about the world around us. However, if we cannot either rationally or empirically be certain of these facts, does it follow that there can be no firm basis for this endeavour? As the seventeenth-century thinkers pointed out, science can only give us *probable* knowledge of the world around us, not *certain* knowledge. This rejection of the idea of absolute scientific fact actually helps us to account for shifts over time in scientific thinking. If we are dealing only with probable knowledge, of course scientific opinion will change over time as the evidential basis for it changes. However, it is important to point out that this in no way undermines the pursuit of science, what it can tell us about the world or the predictions it can make about natural phenomena. Of course, even if we don't have certain knowledge in the absolute sense, we still have hypotheses and theories which are justified beliefs as they relate to the evidence – they seek to describe the world as it actually is and therefore we are fairly safe in suggesting that these represent the truth, or that science can allow us to ascertain facts about the universe. And, as we will discuss in the next chapter, neither does it necessarily undermine any claims that science may make to objective knowledge about the world.

In the 2009 Open University annual lecture, when discussing the future of evolutionary science, Professor Steve Jones rightly pointed out that 'the beauty of a theory, any theory, is that it can be killed by an ugly fact. Wouldn't it be wonderful if Darwinism was killed by an ugly fact, and wouldn't it be wonderful if we could find that fact, but I don't think we will.' Dawkins' response to this was adamant: 'It would be wonderful, but it's not going to happen!' This is a telling distinction between these two great men. It is interesting to note that Jones offers an opinion ('I don't *think*') whereas Dawkins presents a statement of fact ('it's *not* going to happen'). Dawkins is clearly aware, as most scientists are, that we deal in theories, not absolute and certain knowledge, and that

187

these are not one and the same. It was interesting, however, that while one man was celebrating the non-dogmatic and receptive nature of science the second appeared, as I saw it, less enthusiastic about it. I felt that this was a very revealing comment and really got to the core of why Dawkins is criticised as dogmatic. However, it is possible that his continuing focus on the science/religion debate, and fear of giving fuel to the creationists, have ironically made him appear more entrenched and apparently dogmatic than he might want to be.

If we admit that scientific understanding may change over time, this does not necessarily undermine science. The fear of saying things publicly that could be misinterpreted and give succour to creationists is in danger of stifling academic debate. Scientists should and do embrace new cogent and coherent ideas, not fear them. It is perfectly feasible, and acceptable within the model of knowledge that science adopts, that in 100 years the theory of evolution will have changed in ways we cannot possibly even imagine today. And the perceived 'threat' of creationism should not stop us from saying that.

Empirical science is an exciting, objective and open-ended endeavour that can give important insights into how the world works, and its application is clearly of benefit to society. Its representation, direction and communication, however, whether purposely or inadvertently, can be influenced by personal politics or beliefs – for instance, belief in abstract 'truths' such as whether or not there is a God.

The real problem is that Dawkins discusses rationalism as an area of knowledge that goes hand in glove with empiricism. This obviously garners a lot of criticism from those who see the two as separate perspectives on the world with different agendas. In *A Devil's Chaplain*, during one of his typically spirited tirades against cultural relativism, Dawkins challenges 'truth hecklers' – those who argue that science cannot provide absolute truths, and take one step further than I have to suggest that science is a belief system. After lampooning cultural relativism he moves on to philosophers of science, stating that 'a different type of truth heckler prefers to drop the name of Karl Popper or (more fashionably) Thomas Kuhn'.* Then, in typical Dawkins style, he gives a potted account of his imagined adversary's stance:

> There is no absolute truth, your scientific truths are merely hypotheses that have so far failed to be falsified, destined to be superseded. At worst, after the next scientific revolution, today's 'truths' will seem quaint and absurd, if not actually false. The best you scientists can hope for is a series of approximations which progressively reduce errors but never eliminate them.

The rest of this chapter, then, will explore what Dawkins was referring to when he mentioned Popper and falsification, and in the next chapter we will look at the implications of the work of Thomas Kuhn for debates around cultural relativism. In order to understand the issues that some people have with some of Dawkins' assertions about what we can say using science, a bit of background knowledge about the philosophy of science is necessary. This is only meant as an introductory snapshot of abundant work being done in the philosophy of science, but it gives an

---

* Interestingly, he also argues in *A Devil's Chaplain* that this stance is down to the fact that 'philosophers of science are traditionally obsessed with one piece of scientific history; the comparison between Newton's and Einstein's theories of gravitation'. This is a fairly outdated and very narrow perspective on the history and philosophy of science.

insight into Dawkins' stance on science and what it realistically can and cannot say. While clearly Dawkins is no philosopher and nor should we expect him to be, when he has made assertions about the nature of science or on the issues surrounding relativism, suffice it to say that he has not given a fair or accurate representation of the wider debates within academia.

Induction

As we have seen, science is based on empirical accounts of the world around us, and uses observation and experiment to describe natural phenomena. However, this kind of description is only one aspect of the scientific endeavour – it offers us theories about the nature of the world, and data from previous scientific experiments can be used to predict what will happen in the future.

Can it really be that simple? In order to go some way towards understanding when we are justified in thinking that something is scientifically accurate, we first need to understand both what it means to 'test' a scientific theory, and how theories are evaluated based on the results of such tests. Again, this is a complex area with much debate – different philosophers of science have many different perspectives on what science is, what it can realistically tell us about the world, what its methodology is and even what a theory might be. But we will stick to exploring those philosophers whom Dawkins has chosen to tackle – Popper and Kuhn. To explore the work of Karl Popper we need to look at three closely linked ideas – induction, verification and falsification.

As we discussed in chapter 1 the concept of induction was at the heart of Francis Bacon's criticisms of Aristotelian thought. To recap briefly, induction is the term used to describe any process whereby one infers a generalisation or theory from a series of observations. The following classic example is familiar to philosophy students everywhere:

> Observation: All the observed swans are white
> Therefore
> Conclusion: All swans are white

The above is a form of inductive reasoning. We are basing the conclusion on the accumulation of observations on the nature of swans. A slightly different approach is to think about what we can know about the nature of swans deductively. Deduction is based on the use of logic. For example:

> Premise 1: All swans are birds
> Premise 2: All birds are living creatures
> Conclusion: Therefore all swans are living creatures

If the premises are true then the conclusion must (logically) be true. Therefore we logically deduce the conclusion from the premises. If we go back to our earlier example of inductive reasoning:

> Observation: All the observed swans are white
> Therefore
> Conclusion: All swans are white

We can see that inductive reasoning is not like deductive reasoning, insofar as the conclusion does not necessarily follow from the premise or observation. The observation 'all the observed swans are white' only provides inconclusive *evidence* for the conclusion. The observation could be thought to make the conclusion 'all swans are white' more probable, but not certain. What happens when we see a swan that is not white? The swan species that are native to Europe and Asia are all white; only in Australia or South America are there swans that are either black or have black necks. Up until the seventeenth and eighteenth centuries, no European had ever seen a black swan; so before this point it was assumed that all swans were indeed white.*

Scientific investigation can be – and has historically been thought to be – a type of inductive process, where we increase the observational or evidential basis for or against a particular theory without the evidence conclusively establishing the theory. But this leaves us at a bit of an impasse, because we cannot observe everything or every instance of a phenomenon. Therefore, we can never establish once and for all that all swans are white. The problems engendered by reliance on induction can be explored by comparing the work of two philosophers, Rudolf Carnap and Karl Popper.

Verification: Carnap

Rudolf Carnap was a philosopher who was a proponent of logical positivism – a view at the extreme end of the empiricist tradition that rejects all forms of metaphysics – in the early twentieth century, and part of what is known as the Vienna circle.† In short, he

---

* It was actually a colleague of Sir Joseph Banks, the naturalist and ornithologist John Latham, who first described an Australian black swan in the 1780s or 1790s. They had been spotted in 1788 on a voyage to New South Wales by the surgeon John White. It is thought that Dutch sailors may have previously come across Australian black swans in the seventeenth century.

† A group of philosophers of science who met to discuss their perspectives on empirical philosophy between 1922 and 1938. The group was made up of philosophers, mathematicians, social scientists and physicists. The principal consensus that they reached was a form of logical positivism, or logical empiricism, that rejected metaphysics as a path to knowledge. This is for the most part

advocated a method of evaluating scientific ideas through the use of repeated experimental tests.

If an idea is subjected to a series of experiments or tests there are two possible outcomes:

1. Eventually some of the experimental evidence will contradict what we would expect to see if the idea were correct. If this happens, a scientist can reasonably claim that the idea is inaccurate.
2. The results of the series of experiments and tests support the idea.

Carnap suggested that once an idea has passed a sufficient number and variety of tests, it can reasonably be assumed to have been confirmed. However, this confirmation will never give us absolute or certain *knowledge* in any sense. This is partly because to test an idea to the extent that we can arrive at certainty would involve putting together a potentially infinite amount of evidence to show that the scientific statement ('all swans are white') is correct for all possible events that it describes. The most we can say is that a statement is very likely to be true on the basis of the extensive evidence offered by the series of tests and experiments. This evidence means that we can be fairly confident in our predictions about what will happen in the future. Thus we can 'verify' a statement by subjecting it to a variety of tests; and this method

discounted as implausible today, and was widely rejected by later philosophers including Wittgenstein, Kuhn and Popper – for example, one of Kuhn's key criticisms was that this stance tends to be ahistorical and asocial in its approach to understanding scientific knowledge (see chapter 8). Nonetheless, the work of the Vienna circle had a lasting impact on the development of analytical philosophy of science, and while it was perhaps not the best model for scientific practice, it played an influential role in the history of philosophy of science.

is sometimes called verification. The ever-increasing amount of evidence from ongoing experiments or tests will positively verify a theory. Carnap used this method to argue that we can demonstrate a type of truth or probable truth about different kinds of universal statements about nature.

To go back to our swan example: if we continue to observe more and more white swans, it increases the probability that our theory that 'all swans are white' is correct.

There are of course some problems with this verified statement about the universal nature of swans. First, we cannot possibly observe all swans that have ever existed or will ever exist. Even if we *had* in fact seen every single swan, there would be no way to be sure that we had indeed seen them all. An issue that arises out of the verification of a theory or universal statement about nature is that we cannot observe the infinite. To explore a second problem, let us go back to our earlier, apparently verified theory that all swans are actually white. What, then, would happen if we saw a black swan? Under the inductive, or verification, model of science we would have to abandon our theory that 'all swans are white', as we would no longer be able to verify it. But where would this leave us – can we say anything about the nature of swans, or have we failed in our attempt to understand the world around us?

Falsification: Popper

Karl Popper was another of the pre-eminent philosophers of science of the twentieth century. Although Popper was born in Vienna and lived there during the first part of the twentieth century he was not invited to join the Vienna circle, even though some of its members, including Carnap, were aware of his work.

Popper later rejected Carnap's version of verification altogether. His most influential work, *Logik der Forschung* (*The Logic* *of Scientific Discovery*), was originally published in 1935. In this he unleashed his damming critic of the logical positivists.* He thought that any attempt to use experimental evidence to verify or positively demonstrate the truth or probability of a universal statement about nature was doomed to fail from the outset. Popper argued that not only Carnap's inductive logic, but also the very idea of inductive logic, was fundamentally mistaken.

Popper disputed that a theory can ever be truly confirmed through a process of induction, as we can never hope to conduct all the tests that would be necessary. However, we can easily discount or disprove an idea. We can show that a hypothesis is false by discovering one repeatable test whose results contradict it. To illustrate this, we shall return to our swan example.

If we wanted to research the colour of swans, we might start out with the theory that 'all swans are white' based on the limited number of observations we have made previously. Armed with our theory, we go out into the field and observe a large number of white swans – using inductive reasoning, we could happily state that our theory is very probably correct. What then if we decide to undertake our research in Australia and – calamity – we discover a black swan? Under the inductive model of science we should now abandon our theory that 'all swans are white' as incorrect – the evidence does not verify it. We have wasted hours of research, not to mention all that money on a plane ticket to Australia!

The importance of being false

However, Popper's argument suggests that all is not lost. We gain something from this series of tests and experiments – which is that we can now *falsify our hypothesis* that 'all swans are white'.

---

* The translation into English of *Logik der Forschung* was first published in 1959.

What, then, can we learn from our observations of both white *and* black swans? One thing we can definitely support is the theory that 'not all swans are white':

Hypothesis: All swans are white
Observation: The observed swans are white and black
Therefore
Conclusion: Not all swans are white

Popper argued that a scientific hypothesis should attempt to show how certain types of event are prohibited or *falsifiable*. This gives us a much stronger basis for our scientific knowledge than verification does. According to Popper, the process that scientists should follow is one that involves repeatedly subjecting a theory to a series of attempts to falsify it; as a theory successfully passes more and more tests, we can say it has been corroborated.

However, according to Popper, when we corroborate a theory in the sciences this does not necessarily mean that it has a high degree of probability of being correct. It still may be improbable, given the experimental evidence we have collected. We only tentatively accept these scientific theories, while continuing to try to refute them. We use them only until either the theory is eventually falsified or a better one comes along.

A matter of fact
A central concern of Dawkins' later works is the rejection of religion as a scientific claim. But is it really a scientific claim? What is it that makes science different from other ways of understanding the world around us or, to put it another way, what demarcates science from the arts or religion? The *criterion of demarcation* for Popper was that theories which are not refutable, or in other words

falsifiable, by some possible observation are not scientific. Popper argued that science advances by a process of repeated conjecture and falsification. We may be able to corroborate the conjecture that there is no interventionist God, but we may have difficulty in corroborating the conjecture that there is probably no deistic God.* Therefore, as good Popperian scientists we would have to concede that this may be an area on which we cannot definitively say that we have scientific grounds to reject such a claim. Indeed, we may have to accept that any use of reason cannot prove that something does not exist.

In his book *Conjectures and Refutations*, Popper described how he sought to distinguish between science (in this case Einstein's theory of relativity) and religious, metaphysical or pseudo-scientific statements (the examples he used were Marxist history, the work of Sigmund Freud, Alfred Adler's psychoanalytical and psychological approaches and astrology). Here Popper pointed out that the proponents of those theories which he classed as pseudo-science confirmed their ideas through a process of veri-fication which was based on previous observation. He gave an example of how this might not be the best basis for confirming a theory.

Popper related how he had described a child's case to the psy-chologist Adler. Adler responded that because of his 'thousand fold experience' he could analyse the case without ever meeting the child. Popper observed that 'his [Adler's] previous observa-tions may not have been much sounder than this new one; that

---

* By an 'interventionist God' I mean a God that intervenes in our daily life or the natural world directly. By 'deistic God' I mean a God of first cause – for example, a divinity that might oversee a world that works entirely and com-pletely by natural, mechanical laws and processes, even though these laws are said to have been divinely created.

each in its turn had been interpreted in the light of "previous experience", and at the same time counted as additional confirmation. What, I asked myself, did it confirm? No more than that a case could be interpreted in the light of the theory. But this means very little, I reflected, since every example could be interpreted in the light of Adler's theory, or equally Freud's.' He continued:

> I may illustrate this by two very different examples of human behaviour: that of a man who pushes a child into the water with the intention of drowning it; and that of a man who sacrifices his life in an attempt to save the child. Each of these cases can be explained with equal ease in the Freudian and Adlerian terms. According to Freud the first man suffered from repression (say of some component of his Oedipus complex), while the second man had achieved sublimation. According to Adler the first man suffered from feelings of inferiority (producing perhaps the need to prove to himself that he dared to commit some crime), and so did the second man (whose need was to prove to himself that he dared to rescue the child). I could not think of any human behaviour which could not be interpreted in terms of either theory. It was precisely this fact – that they always fitted, that they were always confirmed – which in the eyes of their admirers constituted the strongest argument in favour of these theories. It began to dawn on me that this apparent strength was in fact their weakness.

He later pointed out that the 'two psycho-analytical theories' were 'simply non-testable, irrefutable. This does not mean that Freud and Adler were not seeing certain things correctly … But it does mean that those "clinical observations" which analysts naively believe confirm their theory cannot do this any more than the daily confirmations which astrologers find in their practice.'

Popper then went on to discount the predictions of astrologers, which are 'sufficiently vague' to 'explain away anything that might have been a refutation'. This, he said, is 'a typical soothsayer's trick, to predict things so vaguely that the predictions can hardly fail: they become irrefutable'. On the other hand the predictions that Einstein's theory makes, he argued, are refutable – they are falsifiable and this is the criterion that defines the theory as scientific.

As some readers will no doubt be aware, Popper did at one point in his career, in 1976, suggest that the theory of natural selection was in parts not falsifiable and that it was a 'metaphysical research programme'. This is sometimes used by those who seek to undermine natural selection to argue that it is not a scientific theory. It is important to note, however, that Popper later retracted this assertion. In an article published in the journal *Dialectica* in 1978, he wrote:

> The fact that the theory of natural selection is difficult to test has led some people, anti-Darwinists and even some great Darwinists, to claim that it is a tautology. A tautology like 'All tables are tables' is not, of course, testable; nor has it any explanatory power. It is therefore most surprising to hear that some of the greatest contemporary Darwinists themselves formulate the theory in such a way that it amounts to the tautology that those organisms that leave most offspring leave most offspring.
>
> I have in the past described the theory [of natural selection] as 'almost tautological', and I have tried to explain how the theory of natural selection could be untestable (as is a tautology) and yet of great scientific interest. My solu-

tion was that the doctrine of natural selection is a most successful metaphysical research programme. It raises detailed problems in many fields, and it tells us what we would expect of an acceptable solution of these problems.

I still believe that natural selection works this way as a research programme. Nevertheless, I have changed my mind about the testability and logical status of the theory of natural selection; and I am glad to have an opportunity to make a recantation. My recantation may, I hope, contribute a little to the understanding of the status of natural selection.

A metaphor for life

How then do Dawkins' theories relate to our understanding of science – are they based on falsifiable experimental evidence? A question we will return to in chapter 9 is whether or not the extended phenotype hypothesis has been the subject of concerted research programmes. However, when it comes to the selfish gene approach, Dawkins explicitly states that he is offering a new interpretation of existing data. So it does accord with some experimental data, in that it is designed to be a means of interpreting this data. In terms of the selfish gene concept it is hard to challenge Dawkins on empirical grounds – with the exception of the idea that the immortal gene or replicator is the only thing that is inherited, and that there is no feedback from its environment. Really the debate is about how best to understand and interpret the data – do genes or replicators control their vehicles, or do they interact as equal partners with everything else in the developmental matrix? It's about what offers the best explanation and what opens up new research avenues.

The selfish gene concept has certainly captured the public imagination, but is it simply a metaphor? It is a combination of two things: an empirical claim that only genes or replicators are inherited and are in essence immortal, and an interpretation of evolutionary biology based on this. We can test the first claim, and if we were to find it wanting that would obviously diminish the second. But we can't directly test the second claim because it is not an empirical claim as such – rather it is an organising principle for research, like many other metaphors in science. In many ways the selfish gene approach is meant to be a way to interpret data as much as anything else – metaphors are often used in this way, and it is not in itself a problem. Where it becomes problematic is when the interpretation it offers is not as productive as alternative analyses.

Where does this leave us in our debate about whether or not Dawkins is right to assert that it is a scientific 'fact' that there is no God? I will leave it to the reader to decide how this statement relates to their own beliefs or understanding of religion, or indeed science. Suffice it to say that by Popper's model not only can we not assert that there is no deistic God scientifically, but we cannot even say that there is *probably* no God. We may be able to argue that there is probably no God from a rationalist perspective – and even here there is room for doubt – but this is a very different thing. Dawkins' big mistake in some people's eyes is to claim a rationalist argument as a kind of scientific reasoning or scientific fact. The claim that there is probably no 'first cause' God is simply not a falsifiable scientific fact, and to claim otherwise could act to severely undermine not only the valid practice of science, but also the communication of its findings.

CHAPTER 8

IT'S NOT ALL RELATIVE

*A new scientific truth does not triumph by convincing its
opponents and making them see the light, but rather because its
opponents eventually die, and a new generation grows up that
is familiar with it.*

Max Planck

Apparently then, as far as Dawkins is concerned science is the
wholly objective handmaiden of truth – a truth that transcends society, culture and to some extent history. However, during the last half-century this view has been criticised by some
historians and philosophers of science as being rather a naïve
one that ignores how scientists, and those who study the sciences,
actually perceive the activity 'science' as working. The majority of
scientists recognise that their theories will change over time, without needing to reject any claim to objective knowledge. Science is
a dynamic process, not a static one. Theories that were valuable
tools 30 years ago will have been tinkered with and adapted as
new research data became available. They may even be rejected
if this new data fundamentally contradicts them. One thing that
historians of science learn very quickly is that even individual scientists quite often radically change their own theories over time
– sometimes even rejecting key tenets of their earlier published
ideas.

This raises a number of questions about how science works
and how it is different from other ways of trying to understand
the natural world. If we acknowledge, as most people do today,

that while there is an element of objectivity in science society nevertheless plays a role in its development and direction, this leaves us with some questions about how to define a scientific fact, or indeed 'truth' itself. This brings us right back to Dawkins' pet hate, 'cultural relativism'. In order to understand what this debate is really about, it will be helpful to briefly look at some of the philosophical theories from which these debates spring. While I am certainly not arguing in support of a culturally relativist position, I would argue that here again Dawkins only gives a partial picture of the state of play in academic circles. He has apparently latched on to what in many ways is an outdated version of the debates surrounding the interplay between science, the practice of science and society. What he seems to see as a closed and static debate has actually developed significantly since its heyday in the 1990s. In light of this, Dawkins' response to 'relativist' critics could be damaging to the wider representation of science and its role in society.

Cultural relativism

As we discussed earlier, Dawkins rejects all forms of relativism*
– in terms of both science and ethics. It is a theme to which he returns throughout his work and in his public talks. For example, according to a report on the event by the *Guardian* science correspondent James Randerson, Dawkins asserted in a talk at the 2007 Hay festival that alongside the threat from the religious right

---

* Relativism, in its simplest form, argues that the way we perceive the world or construct our accounts of the world – be they theories, truth claims or our understanding of morality – is relative to or contingent on our social or cultural context. Cultural relativism holds that there is no objective way to evaluate culture – different cultures can only be understood on their own terms, through their own practices or customs.

and creationists, scientists and the rationalist movement 'face an equal but much more sinister challenge from the left, in the shape of cultural relativism – the view that scientific truth is only one kind of truth and it is not to be especially privileged'. Fighting talk indeed, but this is born of a wider academic spat so the overly emotive language is not all Dawkins' own.

This conflict between certain groups of social scientists and scientists is often described by a term coined for an ongoing debate during the 1990s – the 'science wars'. The point of this debate was to ask whether or not 'science' is merely a social construct – if it is wholly based on the subjective opinion or perspective of a scientist, how can it make any claims about objective truths? This was played out in a short-lived series of debates between scientists and science spokesmen, and those whom they saw as 'anti-scientific' relativists or social constructionists.*

The zenith of this spat is often thought to have been the publication of a paper in 1996 by the physicist Alan Sokal in the journal *Social Text*. This paper was a nonsense hoax scientific paper on quantum physics that was supposedly meant to support the postmodern† criticism of objectivity in science. Much was made of the fact that the editors of the journal had so little understanding of the science that they were purportedly analysing that they did not detect the hoax paper and published it. The Sokal paper brought this debate out from academia into the media. There was subsequently a lot of mud-slinging on either side, between those

---

* Social constructionism in its simplest form argues that science has no claim to objective knowledge, as knowledge and experience originate within or are cultivated by a certain society or social group. Therefore, according to this idea, all scientific theories are social constructs.

† Postmodernism is defined here as a rejection of theory or ideology in favour of pluralistic accounts of values and methodological approaches.

who were seen as warriors of science and those who sought to defend the sociological study of scientific knowledge (SSK). The upshot was that the debate became even more polarised and, in the media at least, based on caricatures of the opposing groups' positions. Tellingly, the protagonists on both sides have subsequently expressed regret over the escalation and grandstanding that ensued.

These debates did not just come out of the blue – they were born of areas of research that had been developing since the 1970s, some of which built on the work of Thomas Kuhn about how science progresses over time.

Time will tell

Popper is often quoted as having said that 'science is a history of corrected mistakes'. It appears that this can apply to any arena of academic research. How then does science proceed over time? Clearly, if 'science' is an open-ended endeavour it cannot be seen as simplistically and exponentially progressing towards a predetermined ultimate goal. With the growth in history of science as a discipline throughout the twentieth century, we are now in a much better position to look at how science has *actually* operated.

One of the criticisms of the logical positivists was not only that they were asocial in their approach but that they were ahistorical in their analysis of 'science'. If, it is argued, we remove 'science' from its historical and social context, we cannot really understand how the practice of science works. The key theorist whose work overthrew this perspective was Thomas S. Kuhn. As Dawkins has highlighted, Kuhn is best known for his work on scientific revolutions. Kuhn's work, as well as Popper's, is relatively widely known, not only within the humanities but also in the sciences – though it is important to note that Kuhn did not agree with Popper's idea that science proceeds through a process of falsification.

There is a common misunderstanding of Kuhn's work as feeding into a relativist account of knowledge. As we will see shortly, this is not necessarily the case. Another implication from Kuhn's work that may not sit happily with those who take a logically positivistic approach to the sciences is that theories that were once accepted by the scientific community might be replaced wholesale by completely new ones that may contradict them. This raises big problems which are still the subject of much debate – if theories can change over time, how do we account for any claim to knowledge or truth that current theories may have over previous ones, or indeed future ones? As discussed in the previous chapter, Dawkins picked up on this in *A Devil's Chaplain*, caricaturing philosophical hecklers as saying that 'at worst, after the next scientific revolution, today's "truths" will seem quaint and absurd, if not actually false'.

And while it would of course be patently absurd to suggest that scientific ideas have not changed over time, the real question is whether they progress toward a greater 'truth'. Dawkins, I imagine, would argue that theories in the past were attempts to get it right. The testing that science requires meant that some ideas were abandoned, while others were adjusted to fit with new data. As the theories changed in this way, they got closer and closer to reflecting the world as it is. Does Kuhn challenge this version of the history of science and the concept of progress in science – and is Dawkins right to see his work as a potential philosophical heckle?

Revolutions in science

Thomas S. Kuhn was a highly influential American historian and philosopher of science, who had originally gained a PhD in physics. His most well-known and influential book was *The Structure of Scientific Revolutions*, published in 1962. This has been called one of the most influential books of the twentieth century, selling well over a million copies. While there are a number of criticisms of this text, it has almost certainly led to a near-universal reappraisal of the way we understand and view both scientific knowledge and its history. In *The Structure of Scientific Revolutions* Kuhn suggested a more 'revolutionary' model of the development of scientific knowledge, which is perhaps best explained by discussing some of his earlier works. In his work prior to *The Structure of Scientific Revolutions*, Kuhn had suggested that there may be two distinct types of scientific thought, that which is convergent (fundamentally conservative and traditionalist) and that which is divergent (fundamentally innovative). He further suggested that these two distinctly opposite types of scientific thought are 'inevitably in conflict' – thereby giving rise to an 'essential tension' at the heart of the scientific enterprise. What Kuhn was suggesting was that science progresses not only by the accumulation of knowledge through new observational and therefore empirical findings, as some had previously supposed, but also by the force of intellectual upheaval.

This conception of the history of science was continued in Kuhn's *The Structure of Scientific Revolutions*. The basic premise of this book was that instead of the two types of scientist – divergent and convergent – being in continual conflict, they both formed the basis for different kinds or periods of science. The traditionalist convergent schools of thought, he argued, are responsible for periods of science that are tradition-bound mopping-up operations. These are times when there is a fairly broad consensus about the overarching theories or conceptual frameworks in a scientific field, and Kuhn described this phase as a period of 'normal science'. He then argued that these times of normal science are punctuated by periods that are characterised by more divergent and innovative scientific thinking, or 'revolutionary science'.

Kuhn calls the overarching theories or conceptual frameworks of a scientific discipline 'paradigms'. To summarise Kuhn's paradigmatic historical model of science, he posited that there are certain phases in the development of a scientific discipline:

1. **A fact-gathering phase** – this is before there is a central paradigm, or central organising idea, that acts to unite a group of scientists. There may be people collating data about certain phenomena, but there may also be wildly differing opinions about what these phenomena show. There is no defining paradigm at this stage.

2. **A paradigm emerges** – this is when the texts are published, by more than one scientist, which really define the new discipline or field. These texts would typically outline the methods and approaches to classifying and analysing the natural phenomena that are seen as best practice for future research.

3. **Normal science** – a paradigm is now in place. The researchers in this discipline now go about collecting further evidence to demonstrate the paradigm. They use the paradigm as a predictive model and work out new applications for it. This is not to say that there is dogmatic adherence to the paradigm – it is tinkered with as new researchers enter a field and fresh data are uncovered.

4. **A crisis stage** – this period is precipitated by the discovery of natural phenomena that are so anomalous to the existing

paradigm that it has to be revised. These phenomena might not be explained by the existing paradigm at all.

5. **A new paradigm** – there may be a fairly abrupt transitional phase during which an individual or small group of scientists puts forward a new conceptual and methodological framework. This is called a paradigm shift. This is where, for example, younger scientists new to the field might put forward an innovative or divergent paradigm. Kuhn suggested that this new way of looking at the data may be *incommensurable* with the existing paradigm. The old guard of a discipline might simply refuse to accept the innovative young scientists' new way of thinking – though this is certainly not always the case. Eventually, the new paradigm will gain support and a new period of normal science will ensue.

Incommensurability as Kuhn described it can be seen as a kind of Gestalt switch, that is, the observed evidence can appear different to each group – the traditional old-school scientists or the innovative new researchers – because their perspective is influenced by the theories behind their work. The classic picture used to describe this is the Duck Rabbit Illusion. From one angle the image appears to be a duck and from the other a rabbit. The example that Dawkins tends to use to describe differing opinions within scientific communities is a Necker cube – a simple line drawing of a cube, the front and back of which can be seen from different perspectives. The point here is that you can never perceive both at the same time. A classic example of this kind of incommensurability is William Thomson (Lord Kelvin). As outlined in earlier chapters, he challenged Darwin's theories on the grounds that by his calculations the laws of thermodynamics would not allow for the extended time period Darwin needed for natural selection to occur. Kelvin lived to see the discovery of radioactivity in 1903, which fundamentally changed the general understanding of the age of the earth. However Kelvin, who was fully aware of the new research and its implications, never really accepted them and died four years later without ever retracting his earlier position.

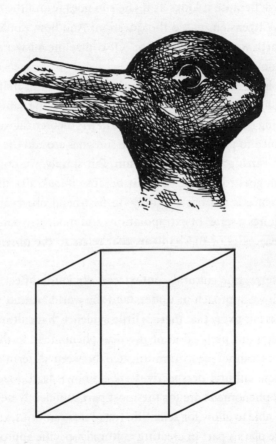

The Duck Rabbit Illusion (top) and Necker cube (bottom).

The third way?

A famous anecdote about Wittgenstein illustrates this quite well. Wittgenstein apparently once asked one of his students why people would ever have thought that the sun went round the earth, rather than the other way round. The pupil reportedly answered 'Because it looks as if the sun goes round the earth,' to which Wittgenstein posed the question: 'And how would it look if the earth went round the sun?' Of course the answer is that it would look exactly the same.

The issue arising from this is that because thinkers past and present have acknowledged that it would look the same, this is something that is taken into account in the theoretical works that have aimed to prove either that the sun goes around the earth or that the earth goes around the sun. Put simply, we don't think the earth goes round the sun just because it *looks* like this is the case. Science is not based purely on first-hand observation, but also requires a series of extrapolations and models to confirm the causal processes or mechanisms that relate to the observational data.

However, our cultural context can obviously affect the way in which we approach or understand the world around us. But I think it is fair to say that there is little evidence that culture or language alter our perception of physical phenomena to the extent that ideas cannot be communicated between differing groups once these cultural perspectives are acknowledged. Macro-level physical phenomena are for the most part sufficiently neutral for us to be able to allow for scientific tests across cultures, and while language plays a part in creating culturally specific approaches to the physical world, these are not insurmountable barriers.

I do, however, think it is also fair to say that cultural, political, economic, linguistic and sociological concerns can act as constraints on open-ended research programmes in that they affect the practice of science. A good example of this is the recent spate of closures of chemistry departments in British universities as they are seen as economically unviable. This is something which has rightly caused a public outcry. We have to acknowledge that both historically and in the modern world, science can be directed by society and ideology in that what gets funded or researched is partly dependent on the social context of the researcher and their institute. There are certain areas of the sciences, with applications ranging from medical to military, that are more likely to get funded and researched than others. There are trends in research science, as in all academic disciplines, and to deny that is untenable.

Rather than a pitched battle between science and left-wing 'cultural relativists', as Dawkins would claim, there are a number of philosophers who are comfortable with both scientific realist accounts of knowledge about the world and the idea that society has an impact on the practice of science. A fairly significant proportion of philosophers of science both seek to unpick the concept of objective reality on which science bases its claim to knowledge, and also question the validity of the social constructivists' assertions. I think the real problem with Dawkins' representation of the philosophy of science is that he presents it as a debate that is closed and polarised, rather than one that has continued consistently within the relevant academic communities, and actually preceded its current 'postmodern' incarnation. Dawkins' perspective on current debates in this area, on both the science and the philosophy, is essentially a very flawed one that buys into media hyperbole rather than engaging with the theoretical arguments themselves. It is fairly widely recognised in the relevant academic circles that the 'science wars' were an inflated misrepresentation of the interplay between philosophy and science that has at times been very counterproductive. To call oneself a relativist these days in some disciplines is seen as tantamount to career suicide!

Truth is rarely pure and never simple

Unfortunately, difficulties arise out of a relativistic approach to scientific knowledge and the concept of truth.

The problem with relativistic approaches in relation to evolutionary science is that they are often used to argue for the teaching of creationism in schools. Such arguments tend to take the line that if scientific *truth* is subject to change over time, as Kuhn argued, then certain theories must be relative to or contingent on their historical period – therefore in different periods there must be different and equally valid models of what is *true*. Incommensurability is taken – by some at least – to imply that we cannot judge whether a change is any better than what came before, so we can't assume that it is a change in the right direction (i.e. towards a better science, or towards truth). It is really a version of cultural relativism through time – in this view, just as some might argue that different cultures have equal claims to the truth today, so then must different historical periods. After all, such people say, the past is another country. Why then should we adhere to the scientific models of today, when science has been based on religious perspectives or arguments from design in the past and may be again in the future? Dawkins' atheistic Darwinism, they might argue, could just be a scientific trend or fashion. This feeds into what can be pseudo-philosophical debates over how science differs, if at all, from other forms of knowledge.

There is no doubt that there are some significant problems with the general understanding of what should be taught as 'science' and what should not. This is not just a problem in the US – as mentioned in previous chapters, there have been some worrying tactics employed by creationists and intelligent design (ID) supporters in Europe. A recent example in the UK is the limited company Truth in Science, who in September 2006 sent a free resource pack to the head of science in every UK secondary school and sixth-form college. According to their own website, the content of these packs 'describes and critiques Darwin's theory of evolution on a scientific basis. It also shows scientific evidence suggesting that the living world is intelligently designed.'* They say that 'the theory of intelligent design holds that certain features of the universe and of living things are best explained by an intelligent cause, not an undirected process such as natural selection.'

Challenging Truth in Science

What is really at stake here is the idea of 'consensus' in science. In the FAQ section of their website, Truth in Science argue that we should 'teach views about origins which are currently held by a minority of scientists, for three reasons'. The second of these reasons appropriates Kuhn's work. According to their website:

> Scientific truth is not determined by consensus. In science the majority can be, and often have been, wrong. There is a great deal of peer pressure in the science community, which can stifle objections to a popular theory. New advances in science often begin with just a few scientists who are prepared to risk questioning the reigning paradigm. Thomas Kuhn explained it like this in his classic *The Structure of Scientific Revolutions*.

---

* http://www.truthinscience.org.uk/site/content/view/109/54/

I would like to take issue with the use of the term 'truth' in this context. No doubt this is exactly the kind of use of the term that Dawkins baulks at as well, but then I would also challenge his rather absolutist version of scientific truth. Dawkins takes on an almost positivistic stance when it comes to the history of science – holding that it is a progressive undertaking, based on the steady accumulation of 'truths' with the occasional correction of past mistakes. This is very apparent in both his historical analysis of science, as we discussed in earlier chapters, and his philosophical analysis of it.

In *A Devil's Chaplain* he argues for a version of absolute truth of which most scientists would stop short, for example stating that 'it is forever true that DNA is a double helix'. A bold statement, but a dangerous one as it is not entirely unfeasible that a non-double-helix structure will be discovered, if it has not been already, occurring naturally.* This is not an open-ended approach to science but a statement of an absolutist personal conviction. The danger here is that he lumps statements like this next to those that are categorically different – for example that if we were to go far enough back in time we would find a common ancestor with a chimp, an octopus or a kangaroo. He conflates scientific *fact* with scientific *theory*. The structure of DNA is a kind of scientific *fact* that can be verified and falsified – it is a description – while common ancestry is a *theory* or hypothesis that explains the observed relationship between species and which can be corroborated by a number of hypotheses across disciplines, of which a number are falsifiable. This does not mean that common ancestry is any less 'true', but just that we use a different type of approach to show

---

* There are, for example, rare occurrences of single-stranded DNA in bacteriophages and some viruses, and apparently all sorts of novel non-helix models of DNA can be produced synthetically.

that it is scientifically valid. There is of course abundant evidence that supports the theory of common ancestry. Dawkins then goes on to state that

> even if they are nominally hypotheses on probation, these statements are true in exactly the same sense as the ordinary truths of everyday life; true in the same sense as it is true that you have a head, and that my desk is wooden. If scientific truth is open to philosophical doubt, it is no more so than common sense truth. Let's at least be even-handed in philosophical heckling.

Unfortunately, Dawkins is making another categorical error here. The statement that his desk is wooden is not the same kind of assertion as that DNA is a double helix or that humans have a common ancestor with an octopus. Saying that his wooden desk is wooden is a tautology – it is logically true but a meaningless statement, a redundant repetition based on the premise that a single observation shows it to be the case, so that single observation is correct.

Truth in Science appear to be arguing for a version of science that seeks to find an absolute kind of truth – that the scientists of today might be wrong and intelligent design will be found to be right is their implication. And ironically enough they use Kuhn to argue that 'scientific truth is not determined by consensus'. Kuhn was arguing exactly the opposite – his model of science within a framework of paradigm shifts is one that progresses through a process of scientific consensus.

We should always be a little wary of anyone who claims to be espousing 'truth', be it as a scientist, a theologian or even – dare I say – an author! After all, what is truth? When we think of something as true we expect it to be a representation of the world as it is.

To say something is a *truth* in a philosophical sense is to say that the subject of the truth statement is true independently of whatever a particular person or group of people may think about it. If, for example, a group of people argued that the earth was flat, it would not make the earth flat. The fact 'the earth is ovoid' would still be true even if we all chose to believe that it was flat. Plato distinguished between knowledge and belief in this way. What is known to be true is true regardless of anyone's opinion. For example, 2 + 2 will always equal 4, and would continue to do so even if we all decided to argue that 2 + 2 = 5.

Some things that we might see as true are more subjective. If I were to say that Marmite is nice, this would not be a statement of 'truth'. I believe Marmite is nice, others don't. My statement about Marmite is a matter of opinion – it is a belief, rather than an absolute fact that can be known by Plato's definition. Marmite will not be either nice or unpleasant regardless of what people say about it; whether it is one or the other is contingent, or dependent, on how we subjectively relate to the taste of Marmite.

Truth-seeking mission

Do natural scientists really seek the truth? In the natural sciences we would expect to assert not a scientific 'truth' but a scientific theory. Evolution by means of natural selection, for example, is a theory and scientific fact. That species change over time is an observable fact, and theories such as natural selection seek to explain the mechanisms of this change. Evolution by means of natural selection remains arguably the most elegant and important of all scientific theories yet devised. But, as we have seen in the previous chapter, in principle the programmes of research that it encompasses have been corroborated through a process of falsification, and it is this that defines it as a theory rather than an absolute *truth* in the philosophical sense. Under a Kuhnian model we might argue that the theory of evolution by means of natural selection has been agreed by a kind of consensus within the scientific community based on research data. It is possible that at some point in the future we might unearth a radical new piece of evidence that blows the theory out of the water. Incredibly unlikely, perhaps, but possible ... in theory.*

We should therefore be wary of Truth in Science when they bandy around assertions like 'Scientific truth is not determined by consensus. In science the majority can be, and often have been, wrong.' It is a gross misrepresentation of what science can and cannot say. It is pure sophistry. They use Kuhn's work in order to support the idea that dissent often involves a subjugated minority of scientists – in their case, proponents of intelligent design against neo-Darwinists. Their website supports this assertion by quoting the following statement from Kuhn's seminal work *The Structure of Scientific Revolutions*:

---

* Again, when we talk about the definition of a theory there is some debate – but, at the risk of gross oversimplification, we would expect a scientific theory to do a number of things:
1. Explain or account for a group of phenomena that occur in the world around us
2. Be based on a hypothesis which has been confirmed through a series of observations or experiments (as highlighted above)
3. Make predictions about what we might expect to happen in future observations or experiments.
Overall, it is a statement about what scientists hold to be the general law, principle or cause of the phenomena which they observe. Science aims not only to describe regularities in the things that scientists observe around them – what is often called observable (or empirical) phenomena – but also to explain those phenomena.

Any new interpretation of nature, whether a discovery or a theory, emerges first in the mind of one or a few individuals. It is they who first learn to see science and the world differently, and their ability to make the transition is facilitated by two circumstances that are not common to most other members of their profession. Invariably their attention has been intensely concentrated upon the crisis-provoking problems; usually, in addition, they are people so young or so new to the crisis-ridden field that practice has committed them less deeply than most of their contemporaries to the world view and rules determined by the old paradigm.

This is a disingenuous use of this quote, as Kuhn suggested that any new paradigm will be based on the interpretation of new anomalous scientific research data. Quite simply, there is no adequate scientific research to support intelligent design so there is unlikely to be a 'paradigm shift' in that direction.

It is easy to see why people like Dawkins might find this view of science and its misapplication as problematic – it could be interpreted as painting a picture of a very subjective, dogmatic and static version of science, which only really moves on with the death of the old guard. Was Kuhn really arguing, as Truth in Science would claim, that peer pressure can suppress certain key ideas or new concepts? There are certainly examples from history of dogmatic power struggles – Cuvier, who was sometimes referred to as the 'dictator of biology', springs to mind – but this does not mean that science as a process is inherently static or inflexible. There is always room for debate or challenges within the scientific community, and at best it is far from dogmatic. It is just that Kuhn's work implies that the idea of progress in science is a bit more complicated than it first appears – it does not necessarily imply a rejection of the concept of progress per se, or imply a form of extreme relativism.

Science in context

One thing that Kuhn's theoretical model for the history of science gives us is the ability to reappraise the role of social, political, religious and institutional influences on the development of science without necessarily reducing the explanatory value of the sciences. In other words, it takes into account the place of the individual scientist within scientific research as a whole. Kuhn's model partly broke down the logical positivists' approach, but this way of understanding the history of science has caused considerable problems. This brings us back to the cultural relativist 'philosophical heckling' that Dawkins refers to. There has been considerable misunderstanding and ill will on the part of some scientists and social scientists, which has as we have seen inadvertently spilled over into the debates on creationism. The irony is that Kuhn himself repudiated some of these relativist approaches that were attributed to him.

As a consequence of the damage sustained by notions of scientific truth during this academic warfare, on certain levels, we may no longer perceive science as the purely objective search for factual information that it was once seen as. Increasingly, scientific knowledge is not just viewed as a matter of gathering new objective *facts*; people also believe that it can be influenced by the cultural and social environment within which scientists work. However, even if we accept the implications of Kuhn's model this does not necessarily mean that we have to downplay the role of science as a way of understanding the world around us. It just suggests that we cannot assume scientific knowledge is based on absolute 'fact' or 'truth', but must concede that it is more like a model of best fit which changes along with our understanding of the world and of the relevance of different scientific theories to us as a society.

However, it would be too simplistic to say that 'science' is just another theory among many other competing theories – religion being one of them. To return to the distinctions outlined earlier, in very basic terms science is founded on empirical observation of natural phenomena in the world around us. Religion is based wholly on personal convictions, systems of belief, and cultural and social factors – not on empirical evidence in the scientific sense. This does not mean that people's religious beliefs are unimportant or should not be respected, but just that they are categorically different from the 'theories' offered by the sciences. Religion cannot provide coherent theories about natural phenomena or make adequate predictions based on previous observation or experimentation – and for the most part does not claim to. Furthermore, young earth creationism and intelligent design are not *nor ever have been* based on empirical scientific evidence or analysis and as such should not be taught as science.

One aspect of 'dissent' in science that Truth in Science and others like them highlight is that there is internal to the scientific community a growing rejection of some of the central tenets of neo-Darwinism. Even if this is the case, it does not mean that this dissent heralds the way for an unfounded, unverifiable or unfalsifiable theory such as intelligent design. It merely means that we might in future look beyond DNA or a purely 'gene-centric' approach in order to understand the natural processes involved in heredity.

It is true that there have been and still are disputes between scientists about exactly how certain mechanisms in the natural world might work. It is also true that there is discussion among scientists about the role of different mechanisms in evolutionary biology – but this doesn't mean that evolution as a mechanistic process does not occur. It simply means we have a lot more research to do – and the beauty of the scientific enterprise is that there is so much more for us to explore in this fascinating and complex universe. If science is an open-ended endeavour in which our understanding of the world changes over time, this does not imply that we can gain no insight into how the natural world works from scientific research – it merely means that we cannot hope to explore the infinite in a human lifetime. There should not be any 'dogmatism' in the sciences. Any scientist worth their salt is reflexive about their understanding of the world – they are open to challenge from their peers and to some extent themselves. Debate and discourse rightly form an active part of any academic discipline.

Dawkins falls foul of the debates around science and relativism for a number of reasons. First, in his response to those who say that science may not be objective he makes assertions that could actually give ammunition to such criticisms. For example, by claiming that the current hypothesis on the structure of DNA will forever be true, he is leaving himself open to inevitable criticism if this is found not to be the case – we now know that sometimes, as we saw earlier, it is not. And by linking this assertion to categorically different claims about the theory of common ancestry he also opens the latter up to unfounded criticism. Second, by saying, as he reportedly did in 2007, that science currently faces a 'sinister challenge ... in the shape of cultural relativism' or dismissing the philosophical hecklers that might cite Kuhn and Popper, he appears to buy into an outdated version of both the philosophy of science and the idea of science in society, particularly in his rejection of the social and historical context of science and its practice. The concept of scientific objectivity and an understanding of the practice of science in its social context are no longer seen as mutually exclusive as they were by some in the 1990s. Such a view is unproductive for a number of reasons, not least because it results in a very politicised and personal perspective not only on the humanities but also on science itself. This runs the risk of giving a representation of the science community – as dogmatic, neo-positivistic and non-reflexive – that no doubt many practising scientists would not necessarily recognise and might even abhor.

CHAPTER 9

THE SWORD OF RHETORIC

Out of our quarrels with others we make rhetoric.

William Butler Yeats

In chapter 9 of *The Extended Phenotype*, Dawkins writes: 'This chapter will be a somewhat miscellaneous one, gathering together the results of my brief and foolhardy incursions into the hinterlands of fields far from my own, molecular and cell biology, immunology and embryology. The brevity I justify on the grounds that greater length would be even more foolhardy. The foolhardiness is less defensible, but may perhaps be forgiven on the grounds that an equally rash earlier raid yielded the germ of an idea which some molecular biologists now take seriously under the name of Selfish DNA.' This could be interpreted as a very humble admission, but one wonders just what Dawkins was trying to achieve by publicly claiming no expertise in areas which he then uses in an attempt to refute some of the most powerful arguments against his selfish gene hypothesis. Dawkins uses a form of advocacy in his writing which cherry-picks from fields beyond his own in order to promote his own worldview. This can sometimes result in him presenting a singular version of what evolutionary theory really represents. So this chapter will aim to establish what he is really advocating – an evolutionary worldview or *his* evolutionary worldview.

Clearly, science communicators do not necessarily have to be active research scientists. Nor should we expect them to be able to give a fully comprehensive account of all the fields in their discipline. Evolutionary biology encompasses a very large number of research avenues. Dawkins is obviously slightly biased towards his previous field – ethology – and it is understandable that he should be on shakier ground with microbiology. But what is problematic is that Dawkins suggests that we should view his work as advocacy of a worldview. This seems to imply that he would seek to impose his perspective of his field as an overarching paradigm – something which should be applied to all areas of knowledge, not just those in his own field or even discipline. Yet his scientific perspective is framed by his understanding of only a selected facet of phenomena studied under the aegis of evolutionary biology. He is then inadvertently imposing a neo-Darwinian perspective born of research in one field onto all the other fields of evolutionary biology, wider academic research and the public perception of them.

Dawkins does fully acknowledge his use of advocacy in some of his works, and recognises that he sometimes gives a biased perspective. In the 1981 preface of *The Extended Phenotype* he gives a heartfelt plea for mitigation: 'I apologize to readers who may find a favourite and relevant work missing from the bibliography. There are those capable of comprehensively and exhaustively surveying the literature of a large field, but I have never been able to understand how they manage it. I know that the examples I have cited are a small subset of those that could have been cited, and are sometimes the writings or recommendations of my friends. If the result appears biased, well, of course it is biased, and I am sorry. I think nearly everybody must be somewhat biased in this kind of way.'

Since his move from zoological research to his last post as a professor of the public understanding of science, he has used this advocate's style to great effect. However, in light of his status as one of the primary public communicators of 'Darwinism' and to some extent the science of biology as a whole, this should perhaps be cause for concern. What then is the end result of Dawkins' passionate advocacy on behalf of evolutionary biology and science – is it to open people's eyes to the wonders of science, or can it entrench positions within and opposition to it?

PUS to PEST

Agendas have to some extent always existed behind the popularisation of science. For example the Society for the Diffusion of Useful Knowledge, founded in 1826, was a Whig political organisation which released a number of cheap and easily available pamphlets and books for the middle and working classes. These focused on scientific topics, and were released in order to counter the output of the more radical penny presses. This kind of 'top-down' version of science communication has rightly begun to be rejected by those working in this field. Popular science communication has more recently tended to lean towards a more interactive model which enables audiences to explore subjects on their own terms.

This shift, which again in some ways has been politically driven, is mirrored in the move from the public understanding of science (PUS) model of science communication, which was favoured in the 1980s and 1990s, towards the active public engagement in science and technology model (PEST). Public understanding of science, which was principally concerned with scientists educating an apparently uninformed lay public, was seen by some as a paternalistic approach which had the potential to worsen the increasing gulf between science and society. It is interesting to note that Dawkins was appointed to his role as professor of the public understanding of science in 1995, when the PUS model was already being criticised. The PEST approach, which is itself not without its critics and failings, seeks to gain active engagement of the public not only in terms of communicating ideas to them, but also in terms of their active involvement in making decisions on science policy.

The popularity of science festivals and other, more interactive social forums of science communication – for example Café Scientifique in the UK* and its antipodean cousin Science in the Pub† – are testament to this kind of public engagement in science. Advocacy plays only a small part in these kinds of forums, which focus more on public engagement and involvement in scientific discourse. The assumption is that the audiences and scientists involved in such dialogue-based events will share their experiences and viewpoints and potentially reconcile them. An important point to remember here is that 'public' includes a vast array of people, from schoolchildren right through to scientists from other disciplines, and even those who themselves work in science communication or even science policy. Any audience can represent a wide range of different perspectives, the discussion of which can be as beneficial to the scientist as to the audience.

The important shift here is that the PUS model of science communication is seen, in very simple terms, as a transfer of the canon of science from an expert to uninformed laypersons. However, at best under a PEST model the object is not to inform or tell an audience what to think, but to allow them to truly engage with the subject matter from their own perspective. It has the potential to allow for a two-way communication process. For instance, on its

---

* www.cafescientifique.org

† www.scienceinthepub.com

website the Café Scientifique network is described as 'not a shop front for science' but a 'forum for debating science issues'.

Dawkins, with his self-confessed use of advocacy and his well-documented fondness for rhetoric, is a science communicator very much in the old-school PUS mould. And given his responses to some of his more recent critics, it is clear that Dawkins sees a vast proportion of his potential audience as uninformed and does not necessarily respect their right, within reason, to their own opinion. Examples abound, from his dismissal of *Guardian* columnist and associate editor Madeleine Bunting as an ignoramus as we discussed in chapter 6 to the following description on his website of one atheist critic: 'I could have overlooked the patronizing condescension of his remark, if only he hadn't sounded so smugly *satisfied* by this lamentable state of affairs.' Those who seek to adopt Gould's non-overlapping magisteria view of evolutionary biology – like the historian and philosopher of science Michael Ruse – are labelled the 'Neville Chamberlain School of Evolutionists' in *The God Delusion*; Neville Chamberlain, of course, being best known for his policy of appeasement directed towards the Nazis before the Second World War.

At the very extreme of what I would regard as inappropriate name-calling, on his website he even recalls one lamentable comment that he made in response to an erstwhile acquaintance with whom he had recently had some spat – out of which I suspect neither man came out looking particularly appealing. Dawkins published his letter of response to his sparring partner Rabbi Shmuley Boteach in which Dawkins argues that he 'did not say you think like Hitler, or hold the same opinions as Hitler, or do terrible things to people like Hitler. Obviously and most emphatically you don't. I said you shriek like Hitler. That is the only point of resemblance, and it is true. You shriek and yell and rant like

Hitler. Not all the time, of course. You also tell very good jokes…'*
I am not implying that Dawkins is in any way anti-Semitic, but I
will let the reader decide if this is an acceptable response to someone who is Jewish, regardless of whether they might advocate
anti-evolutionary views.

The 2007 TV series *The Enemies of Reason*, presented by
Dawkins, exemplifies his 'us and them' attitude to those who may
not share his perspective or those with whom he enters into debate.
Even the title has an inflammatory, almost crusade-like rhetorical
air to it. In the series he tackled various opponents who adhered
to different classes of 'superstitions', from astrology and water div-
ination through to homeopathy. Clearly there are good reasons to
address some of these subjects. Confronting these superstitions,
though, is part of a wider programme of consciousness-raising
about science for Dawkins, and he regards 'the current backlash
against science as a betrayal of the Enlightenment'.

The problem, of course, is that in the programme Dawkins
did not really address the reasons why there might be a 'backlash' against 'science'. After the involvement of the military in so-
called 'big science' from the Second World War onwards including
the American atomic bomb programme, right through to more
recent scandals including the MMR vaccine débacle in the UK, the
sometimes disastrous communication over GM crops and even
the long battle to gain official recognition of climate change, is it
really surprising that there is a lack of public trust in science? The
political and social implications for the applications of science are
too prevalent, in some people's minds, for them to see science as
a wholly objective process. Who are people supposed to believe,

---

* On both: http://richarddawkins.net/article,2591,Richard-Dawkins-
Responds-to-Rabbi-Shmuley-Boteach,Richard-Dawkins
http://www.beliefnet.com/story/233/story_23310_1.html

and how can they learn to discern for themselves between 'good' and 'bad' science, if it is communicated via a one-way street? It is no longer the case that we can just treat the 'public' as a impassive, ignorant mass waiting to be informed of the canonical truth by individual scientists. I personally do not think that we can afford to ignore people's genuine concerns, their lack of trust about certain aspects of 'science' or the lack of clarity on what science can say about or contribute to society. Could it be that the hard-line rhetoric and advocacy that Dawkins employs is part of the problem rather than the solution?

Professor of advocacy?

It is probably best to clarify what is meant in this context by the term 'advocacy'. Dawkins has at some points in his career, perhaps unfairly, been criticised as a kind of 'armchair' scientist – someone who writes about science, but is not actively researching the field about which they are writing. This, if used to refer to the latter part of his career, may be a somewhat accurate description and he has occasionally admitted to this approach. However, Dawkins was from 1995 until 2008 a professor of the public understanding of science – we should not necessarily have expected him to be actively researching evolutionary mechanisms while in this post. I am therefore not seeking to deny that there is a role for armchair-style scientific writing in popular science. However, 'advocacy' is not simply a form of 'armchair science' – if anything, it is more a form of rhetoric. Dawkins is a very good wordsmith, and performs extremely ably in television and radio interviews. He can be very persuasive, and regardless of what he is saying the very forcefulness of his character and the style of his argument are both very compelling. This brings to mind Bacon's criticism that scholars of his time were more concerned with 'the choiceness of the phrase' than the 'soundness of argument'.

Obviously, it is quite difficult to completely avoid advocacy, or arguing publicly for one particular position, in any kind of academic writing. I have no doubt been guilty of this myself. Our personal perspective is always liable to slip through even if we try to be even-handed. When writing, a scientist's disciplinary background or knowledge of certain areas is also likely to give a certain, slightly skewed perspective. Even a textbook on 'Biology', for example, could be said to be advocating one particular approach within the discipline, but if it is to succeed as a textbook it has to give a broad enough overview of the generally accepted wisdom across different groups within the discipline of biology, not just the field in which the author's research is based. In addition, it would be pretty hard to write a textbook on any subject that did not 'advocate' the accepted central organising ideas within that discipline or the accepted emergent aspects of new areas of research. A journal article may also advocate a certain position in relation to the research background or perspective of the author or authors. However, most academic journals have a process of peer review, whereby other academics in the field review and give feedback on every article before publication.

We must remember that, while Dawkins has published many peer-reviewed articles during his career, his popular books do not serve the same purpose as a textbook or a journal article, thus they are not necessarily scrutinised by the academic community in the same way. In addition, they were not necessarily written in order to put across the consensus view of animal evolution or the wider evolutionary sciences. Indeed, *The Selfish Gene* was a bold and successful attempt to popularise the challenges to naïve group selectionist perspectives, borne out by Dawkins' earlier doctoral research. In a way it contributed to the consensus, rather than reported it. And for this reason it undeniably remains an important work. Dawkins had a more than respectable scientific career before he became a full-time science communicator. Before 1995, when he took up his chair in PUS, Dawkins held both research and lectureship posts in zoology, rising to be *ad hominem* Reader in Zoology at Oxford. He was obviously well qualified to report on and communicate research in these areas. However, he is still marketed as a scientist and sometimes even a philosopher, and an inadvertent upshot of this is that one might assume he has been working as an active research scientist for the past fourteen years. However, increasingly, as research in the wider evolutionary sciences moves on he is not necessarily seen as representative of the wider discipline by an increasing number of those who do still actively research in those areas.

Clearly, there are still plenty of scientists who do support Dawkins' views, but conversely there are plenty who do not – and it is generally accepted in some research communities that he does not speak for the discipline. Views are varied; some vehemently defend Dawkins' contribution to both science and the debates about scientific reason. In the middle ground, some scientists I have spoken to appreciate the contribution his earlier works have made, but recoil from his more recent grandstanding on science, rationalism and religion. Some, maybe a little uncharitably, simply don't think of him as a 'scientist' at all. Others see him as part of the old guard, ready to be replaced by new models and ways of understanding evolutionary processes. Only time will tell if this really is the case, but we are certainly in an interesting and exciting period in the biological sciences.

Extended advocacy

In his later works, Dawkins was writing for the wider public and here his use of advocacy may seem less problematic. But he also unashamedly embraced the use of advocacy in his earlier scientific writing, when many others would see it as something to avoid wherever possible. The very first line of Dawkins' most academic text, *The Extended Phenotype*, reads: 'This is a work of unabashed advocacy.' In the 1981 preface, Dawkins himself acknowledged that this advocacy would not necessarily be accepted in a scientific journal. He stated:

> Well I suppose I am just a little abashed! Wilson (1975, pp. 28–29) has rightly castigated the 'advocacy method' in any search for scientific truth, and I have therefore devoted some of my first chapter to a plea of mitigation. I certainly would not want science to adopt the legal system in which professional advocates make the best case they can for a position, even if they believe it to be false. I believe deeply in the view of life that this book advocates, and have done so, at least in part, for a long time, certainly since the time of my first published paper, in which I characterised adaptations as favouring 'the survival of the animal's genes' (Dawkins 1968). This belief ... was the fundamental assumption of my previous book.

As it is not 'a factual position' that Dawkins is advocating, but rather 'a way of seeing facts', he goes on to warn his readers 'not to expect "evidence" in the normal sense of the word'. As we discussed in chapter 7, the concepts he is outlining are metaphors which are meant to be an organising principle for research. Even so, this assertion – that it is reasonable for him to advocate his belief and to not seek to provide evidence – seems strange coming from someone who would later go on, under the guise of science and reason, to vehemently challenge other beliefs, be they in religion or in the need to respect other opinions.

The Extended Phenotype tends to be seen as his most important work. The 'selfish gene' concept is, even by Dawkins' own account, merely an extension and popularisation of existing ideas. In a nutshell, Dawkins' extended phenotype concept argues that the expression of the genes of an individual organism can be extended beyond its direct physical boundary. One of the examples that is often cited is the example of genetic influence over the quality of dam a beaver constructs, or the quality of a nest a bird might build. The variations in the genetic makeup of an organism, it is argued, will correlate with an improved ability to build these external structures. And a beaver who builds a better dam is more likely to have a selective advantage.

How does the science community view this work? According to a press release entitled 'European evolutionary biologists rally behind Richard Dawkins' Extended Phenotype' for a workshop held in Copenhagen in 2008 by the European Science Foundation, there are those who still actively support the extended phenotype concept. While apparently there was some debate at this workshop between Dawkins and those who supported a niche construction perspective, it was argued that the extended phenotype concept is as important today as it was when it was first published.* The

* Niche construction theory suggests that this kind of genetic influence over the organism's environment can extend to a process of altering the selective pressures and therefore goes beyond the processes suggested by the extended phenotype perspective. Some proponents of niche construction, including John Odling-Smee, Kevin Laland and Marcus Feldman, have argued that it should be viewed as second only to natural selection in its importance to the evolutionary process. Dawkins views some aspects of the niche construction

organiser, David Hughes from Harvard University, commented that it was recognised that the extended phenotype had 'explanatory rather than predictive power', and that it was 'a good way of looking at things but not necessarily the best approach when designing experiments'. In an article which was published in the journal *Biology and Philosophy* in 2004, Dawkins himself reflected on the scientific impact and credentials of his extended phenotype concept. This article was a response to three preceding articles which had given a critical retrospective on his book. Dawkins mused on what he would say if invited to give his own retrospective. After acknowledging that, in his words, 'the part of the theory that I am not responsible for inventing', namely the 'gene's-eye view', has moved to the forefront of 'ethologists, behavioural ecologists, sociobiologists and other evolutionary biologists' minds', Dawkins admits that

> the part of the theory which is wholly my own, the extended phenotype itself, unfortunately cannot yet make the same claim. It lurks somewhere near the back of biologists' minds, but not in the lobes that plan research in the field. Twenty-one years ago, I said that nobody had done a genetic study using animal artifacts as the phenotype. I think this is still true. I would admit to disappointment, except that it invites the obvious retort: why don't you get out there and do it yourself, then? It is a fair point. I should. Maybe I will. Idleness is a poor excuse, and a preoccupation with writing books only slightly better.

---

theory as a special case version of the extended phenotype, but rejects the implication that niche construction could extend to modify selective pressures across ecosystems.

Dawkins then outlines his dream of a possible future world in which an 'Extended Phenotypics Institute' is set up to research his theories.* There are two important points here with regard to the debate about advocacy in science communication. First, the gene's-eye view is largely representative of the perspective of certain biological fields – as Dawkins states, 'ethologists, behavioural ecologists, sociobiologists'. However, many would strongly contest the idea that the gene's-eye view is as comprehensively embedded or accepted in other fields within evolutionary biology. Second, if, as Dawkins himself admits, the extended phenotype concept is still not a research theme in any scientific institution, while it is very interesting and thought-provoking, how can his continued advocacy of his own concept in the face of scientific criticism today be classed as anything other than speculative?

A blind advocacy

In the 1986 preface to *The Blind Watchmaker* Dawkins again unashamedly claims his work to be a kind of 'advocacy', which he uses to mean explaining something so that the reader 'feels it in the marrow of his bones'. Dawkins argues that it is simply not enough 'to lay the evidence before the reader in a dispassionate way. You have to become an advocate and use the tricks of the advocate's trade. This book is not a dispassionate scientific treatise.' He later continues: 'Far from being dispassionate, it has to be confessed that in parts this book is written with a passion which, in a

---

* And I am sure that he is encouraged to hear that, excitingly, there were apparently proposals to apply for funding for two such research programmes as an upshot of the Copenhagen workshop. According to the press release, 'A follow-up funding application is a great idea and it should focus on examining the parasite manipulation of host behaviour component', said Hughes. 'There was also a suggestion to have a separate application for a nest construction pan-EU network.'

professional scientific journal, might excite comment. Certainly it seeks to inform, but it also seeks to persuade and even – one can specify aims without presumption – to inspire.' What Dawkins is advocating is not only 'that the Darwinian world-view happens to be true, but that it is the only known theory that could, in principle, solve the mystery of our existence'.

Dawkins then relates how shocked he was to find that university debating societies are used as a testing ground for students' advocatory skills, stating that he 'resolved to decline future invitations from debating societies that encourage insincere advocacy on issues where scientific truth is at stake'. There are obviously only certain subjects that he deems worthy of a rhetorical approach. And it is no surprise that 'Darwinism' is top of the list. Dawkins argues that 'For reasons that are not entirely clear to me, Darwinism seems more in need of advocacy than similarly established truths in other branches of science ... Darwinism, unlike "Einsteinism", seems to be regarded as fair game for critics with any degree of ignorance.' I would have thought the difference was obvious; the clear difference is of course that 'Einsteinism' does not have such potential to affect how we view society. In addition, the theories of special and general relativity have not been represented in quite the same way as Darwinism as being atheistic or as standing in opposition to religion. Needless to say, the common misconception that Darwin said that humans are directly descended from monkeys does not help. It is not that Darwinism needs advocating more than other disciplines; it is that its communication needs to be put into a proper historical context. We cannot ignore the misapprehensions that have sometimes inhibited people from approaching Darwin's theories with an open mind. In this case, I would question whether aggressive advocacy or the use of rhetoric is really likely to help.

I am certainly not arguing that the evolutionary sciences do not need to be better communicated or more widely engaged with than they currently are. I am equally passionate about communicating this area of science and the surrounding debates. However, what I would challenge is whether using advocacy is really the best way to go about it, or whether in this context it is a valid form of science communication. What we really need to ask ourselves is what we should expect from a professor for the public understanding of science. Given the perception that there is decreasing public trust in science and scientists, how can we expect people to regain their trust if each individual scientist aims to advocate a certain perspective? How can the lay person distinguish between personal perspective, a metaphor and scientific consensus, without the access, the necessary background knowledge or the time to go through all the relevant research? To be fair to Dawkins, in some of his works he has given a fairly good account of a particular form of neo-Darwinian evolutionary theory, but in others – especially perhaps his best known though least scientific books, *A Devil's Chaplain* and *The God Delusion* – the reader is subjected more to his personal opinion than to a balanced perspective on evolutionary biology.

This should lead us to wonder what Dawkins really means when he uses the term 'advocacy'. Is it truly the advocacy of one who seeks to promote their own version of a discipline over and above any explanation of the wider received wisdom on it or its central organising concepts? What could possibly motivate him to do this? If Dawkins is promoting a 'world-view' as he claims, we should consider what his agenda might be, and which worldview he seeks to replace.

Readers of *The God Delusion* can be left in no doubt that religious belief is one factor. However, this anti-religious stance is not so much in evidence in Dawkins' earlier scientific writing. So what worldview was he seeking to promote when he claimed in *The Blind Watchmaker* and *The Extended Phenotype* that advocacy was a suitable method of science communication? Clearly one of the things he was trying to discount was a non-Darwinian worldview. However, as we have seen there is more than one type of 'Darwinism', and evolutionary thought today encompasses a lot more than just Darwin's theories. Even in the mid-nineteenth century Darwin was not the only 'evolutionary' thinker. Nor has the neo-Darwinian view always been the predominant way of explaining evolutionary mechanisms. There have, as we explored in earlier chapters, been a number of challenges to the neo-Darwinian theory of evolution. These provide the key to understanding the worldview which Dawkins was trying to displace – namely, any worldview that allows the possibility of challenges to the classical neo-Darwinian crown. Before *The God Delusion* came the non-neo-Darwinism delusion!

Another area in which there has been much discussion of the role of 'advocacy' is the climate, or environmental, sciences. Perhaps the most well-known critic of this kind of 'advocacy' is Bjørn Lomborg, author of *The Skeptical Environmentalist: Measuring the Real State of the World*. While I tend to strongly disagree with the conclusions and implications of this book, its central conclusion – that the extreme, 'dire' predictions of global climate change are not necessarily scientifically sound and that they are promoted by environmental advocacy groups, such as Greenpeace, in order to raise funds – is an intriguing one. This is not a new idea – it is actually the central plot device in Ben Elton's novel *This Other Eden*. In this fictional account the arch-technologists peddling bio-domes for the rich to escape into after a predicted environmental collapse are in cahoots with the environmentalists – both are working together to hype up the catastrophic environmental predictions for their own gain.

One could ask – obviously without implying the same level of volition – if there might be a similar relationship between the popular 'advocacy' of an atheistic version of neo-Darwinism and the perceived threat of creationism. This is certainly evident in the way in which Darwin's theories have been represented over the years since his death. Advocacy of 'Darwinism' is not used merely as a response to the supposed threat of creationism, but also in order to counter the usurping of the Darwinian crown by other scientific theories concerning the mechanisms involved in evolutionary processes – from latter-day saltationism or group selectionism, through to recent incarnations of the inheritance of acquired characteristics. Anyone who challenges 'Darwin' or 'Darwinism' is apparently playing right into the hands of the creationists, as we saw in his response to the *New Scientist* 'Darwin was Wrong' headline. There have, however, been throughout history different interpretations of what the terms 'Darwinism' and 'evolutionism' represent, and at some points these have been strikingly at odds with those we might recognise today. Use of the Darwinian label has long been a way of implying the legitimacy of one perspective over another – a tradition which dates from the earliest neo-Darwinians in the late nineteenth century, and has continued right through to the appropriation of all things Darwinian by the most recent crop of neo-Darwinians.

Anti-Darwinism and group selectionism

As we saw in earlier chapters, it is simply wrong to paint the recent history of evolutionary biology as a static, gene-centric affair. The history of science is a much more complex matter than is often portrayed. It simply wasn't the case that Watson and Crick elucidated the structure of DNA in 1953 and from then on everyone just focused on 'genes'. While, to a member of the general public, it might sometimes appear that Dawkins speaks for all and that the 'selfish gene' concept is the central focus of evolutionary biology, this is quite simply untrue. In the second half of the twentieth century there were many challenges to the focus on 'genes' or DNA as the sole unit of variation, inheritance and selection.

Given his most recent texts, it would seem that Dawkins is most concerned by those who seek to oppose Darwinism from a creationist stance. However, it is not merely those who oppose 'Darwinism' per se whom Dawkins has sought to dismiss. The championing of the neo-Darwinian stance was apparently Dawkins' primary concern in his earlier works. What, then, are the possible motives for Dawkins' advocacy of a particular neo-Darwinian perspective?

Dawkins comes from a tradition within the evolutionary sciences that has historically had little time for non-neo-Darwinian mechanisms of evolution. This has not necessarily been the case in other disciplines, most notably areas like medicine, plant sciences and microbiology. However, within his own sphere, throughout the 1980s and 1990s Dawkins was indeed to some extent promoting a consensus version of mammalian evolution. *The Selfish Gene* may have been a treatise against group selection theories (and to some extent this is still a theme in Dawkins' stance), and there were some who were less than comfortable with some of the implications of a hardline gene-eye view of the world, but his wider selectionist neo-Darwinism perspective was to become quite widely accepted. However, as group selection theories have gradually become less widely discussed,* other theories that challenge

---

* Although as we saw earlier this is beginning to be revisited by some; for example, E.O. Wilson and D.S. Wilson.

Dawkins' selfish gene concept have replaced it. As we explored in earlier chapters, a number of research areas have developed that rely less and less on Dawkins' version of neo-Darwinism. While these are beginning to gain ground today, during the latter part of the twentieth century, any public attempt to dispute aspects of the Dawkins/neo-Darwinian perspective tended to be labelled maverick, Lamarckian or 'anti-Darwinian'. At the very extreme end of things, any such criticism is presented as giving fuel to the creationist camp. In Dawkins' work, it is this arena in which we see his advocational style of writing really come into its own. For instance, the last chapter in *The Blind Watchmaker* is entitled 'Doomed Rivals'. Here he outlined his outright opposition to 'Lamarckian types of theory', which have been in his view been 'traditionally rejected – and rightly so – because no good evidence for them has ever been found (not for want of energetic trying, in some cases by zealots prepared to fake evidence)'.* Dawkins, while stating that 'this is no history book', went on to give a highly skewed and ahistorical potted account of the history of Lamarckism, neo-Lamarckism and the role of the concept of 'inheritance of acquired characters' in evolutionary thought – Darwin's acceptance of the inheritance of acquired characteristics was swiftly dispatched and we were incorrectly told that these were 'not part of his theory of evolution'. Indeed, this is a matter of such little apparent status that Dawkins stated: 'instead of examining the evidence for or against rival theories, I shall adopt a more armchair approach.' Dawkins returned to this rejection of inheritance of acquired characters in *The Extended Phenotype*.

Only Dawkins himself can really give any insight into his continued enthusiasm for promoting neo-Darwinism in the way he

* This is presumably referring to the scandal surrounding Kammerer's midwife toad experiments in the 1920s which were discussed in chapter 4.

does. I am sure that it springs from the best possible motives, and it is not surprising that he might respond negatively to scientific theories that challenge his own. Historically, however, this is not the first time that the status of natural selection as the only important mechanism for evolutionary change has been challenged in this way. In a sense, even though Wallace's and Dawkins' 'neo-Darwinian' texts are separated by nearly 100 years, they are both concerned with the same threats to their versions of 'Darwinism'.

Wallace's *Darwinism* – the birth of anti-Darwinism

Wallace's *Darwinism* was published in 1889. Wallace was (as discussed in chapters 2 and 3) an ardent supporter of natural selection at the level of the individual, and in fact attributed far more to the evolutionary mechanism of natural selection than Darwin himself. It was partly in order to describe Wallace's panselectionism that the terms neo-Darwinism, 'Ultra-Darwinism' and 'Pure Darwinism' were coined in the late nineteenth century In his book *Darwinism After Darwin*, George Romanes railed against the fact that 'those biologists who of late years have been led by Weismann to adopt the opinions of Wallace, represent as anti-Darwinian the opinions of other biologists who still adhere to the unadulterated doctrines of Darwin'. Romanes – who is sometimes referred to as Darwin's chief disciple – had, much to his displeasure, been depicted as 'anti-Darwinian' by Wallace in an article entitled 'Romanes Versus Darwin', published in the *Fortnightly Review* in 1886. Interestingly, it was in this article that Wallace noted the power of being able to link one's theories to Darwin's:

> Dr. Romanes is well known as an authority on some branches of animal physiology and psychology, and is also an earnest student as well as a great admirer of Mr.

Darwin's works; while, as he informs us, he had for many years 'the privilege of discussing the whole philosophy of evolution with Mr. Darwin himself'. His conclusions on this subject are therefore likely to be widely adopted, more especially as the question is a very difficult one, and the value of the arguments adduced can hardly be estimated by persons who are not well acquainted with the copious literature of the subject.

This is as true today as it was in the 1880s. Darwinism still sells a book or an idea today as much as it did directly after Darwin's death. However, Wallace was less ready to have this criticism applied to himself. Wallace was very much promoting his own version of Darwinism, which ignored Darwin's references to other mechanisms of evolutionary change and outright rejected Darwin's theories of pangenesis which allowed for the inheritance of acquired characters. To quote Wallace in his introduction to *Darwinism*: 'Although I maintain, and even enforce, my differences from some of Darwin's views, my whole work tends forcibly to illustrate the overwhelming importance of Natural Selection over all other agencies in the production of new species.' He later continues to say that his book is 'pre-eminently the Darwinian doctrine, and I therefore claim for my book the position of being the advocate of Pure Darwinism'.

It is interesting to note that Dawkins is perhaps unwittingly repeating Wallace's stance by choosing to overlook what some proponents see as the contemporary equivalent of Darwinian pangenesis. At the time that Wallace wrote this passage there was a fairly strong acceptance in scientific circles that mechanisms other than natural selection may play either a primary or a subordinate role in evolutionary processes. This time period has since been called the 'eclipse of Darwinism' – during which there were a number of competing theories which attempted to challenge 'Darwinism'. Contrary to how this is sometimes portrayed today, the challenges with which Wallace was predominantly concerned did not come from religious quarters. One of the key theories which did concern Wallace was any form of the inheritance of acquired characters. So it was not religion, but the threat of a set of theoretical mechanisms which some – such as Herbert Spencer and Romanes – used to suggest that natural selection was not 'all sufficient' as an explanatory evolutionary mechanism, that spurred Wallace's advocacy of 'pure' Darwinism.

Some of these ideas have certainly not disappeared over the years – Dawkins has also had his share of altercations with members of the scientific community who show a willingness to look beyond selective processes. So far the comparison between Wallace and Dawkins is fairly clear. And interestingly enough Dawkins does recently appear to have been arguing for wider acknowledgement of Wallace's role alongside that of Darwin. However, there was one very important aspect of Wallace's thinking that would, I imagine, be anathema to Dawkins. In the final chapter of *Darwinism*, Wallace outlined his controversial theories as to the possible spiritual origins of the human mind.

Wallace was no stranger to the use of advocacy in his writing – he was well known for his popularist writings which ranged from arguments for the systematic nationalisation of land, to the adoption of overtime pay, through to his endorsement of the anti-vaccination movement. The first page of his obituary, published in 1913 in *Science*, recalled that:

To some his frequent advocacy of unpopular causes suggested perfect indifference to public opinion, and a total disregard of ordinary prudence. Whether, in this or that matter, we believe him to have been right or wrong, we must admire a man who always had the courage of his convictions; and so far from being indifferent to the feelings and opinions of others, his sympathetic nature and longing for fellowship caused him to so zealously expound what he believed would be helpful to other men.

It seems that advocacy has been the weapon of choice for both these men of science when challenging perceived threats to their own worldviews, be they in the realm of science or that of society. That said, I imagine that Wallace might perhaps come under the label 'left-wing liberal' in Dawkins' view. It is important to remember that, for Wallace, the acceptance of alternative views to his idea of selectionism was a very real threat to the credibility of his 'pure Darwinism', for his was a period when a number of theories were competing with any form of 'Darwinism' for the evolutionary crown. Wallace, as the celebrated 'co-discoverer' of Darwinian natural selection theory, had a lot to lose in terms of public profile and the status of an important aspect of his life's work if this aspect of evolutionary theory was to become subsidiary or, worse, outright rejected. Yet it is uncertain whether this is also the case for Dawkins – what would happen if his selfish gene metaphor were to be rejected by the mainstream scientific community? Some might argue that it already has been. However, this is all speculation; only time will tell if there is any significant challenge that will eventually supersede his particular selfish gene version of neo-Darwinism. I doubt whether his status with his sometimes ardent supporters would be much shaken. And now, regardless of the acceptance of 'selfish genes' or 'extended phenotypes', his reputation rests not on his scientific ideas but on his high-profile challenges to religion – so his public profile is perhaps forever assured.

To some extent works of advocacy, like those of Wallace and Dawkins, rather than acting to stimulate scientific understanding, research or debate, might inadvertently act to restrict the development of or public engagement with certain fields in evolutionary biology. Wallace certainly left a legacy in terms of the representation of what counted as acceptable under the 'Darwinism' banner. There are a number of very exciting areas of research currently opening up that simply are not aired as much in the public sphere as they could be. And when these ideas *have* been discussed, they are still quite often couched in terms of 'anti-Darwinism', 'Lamarckism' or 'heresy'. Even *New Scientist* fell into this trap by proclaiming that 'Darwin was Wrong' on its front cover. The message this sends out to the research community is that these are no-go or maverick areas of research, and some researchers who were publicly labelled 'Lamarckian' in the 1990s were quick to distance themselves from this, while others embraced its non-conformist undertones.

Dawkins' advocacy of an anti-religious evolutionary 'science' has not only fuelled the feud between the warring parties of 'Darwinism' vs. 'creationism', but also added to the stigma of becoming involved in the long-running 'neo-Darwinism' vs. 'Darwinism/Lamarckism' debate. Wallace used his advocacy of 'pure Darwinism', or neo-Darwinism, to discredit other theories. Dawkins picked up Wallace's banner, and is today the archetypal neo-Darwinian in a world where the term is becoming less relevant every day. Perhaps the early twenty-first century will come to be known as the 'eclipse of neo-Darwinism'. Therefore Dawkins is (rather ironically, considering their respective views on the spiritual realm) continuing in the footsteps of Wallace. This might well turn out to be a successful strategy; after all, Wallace succeeded in promoting a wholly selectionist version of 'Darwinism' which, after a hiatus from the end of the nineteenth century to the beginning of the twentieth century, held sway for most of the later part of the twentieth century. In addition, Darwin's 'Darwinism' is in some ways as different from that of Dawkins as Wallace's – yet Dawkins, as Wallace did before him, apparently claims to be the standard bearer, above all others, for Darwin's legacy. While it is relatively easy for us to analyse the impact of Wallace's promotion of 'pure Darwinism', it is still too early to coherently analyse the impact of Dawkins' advocacy.

One could even ask whether it really matters if an evolutionary scientist is a 'Darwinian' or not. This term, when used as a label, could conceivably be more limiting than liberating. I am sure that there are many scientists, particularly researchers outside the UK, who have long questioned the significance of the adaptationist model and the use of the Darwinian label. In the US the term 'Darwinism' is often exchanged for 'evolutionary science', because the former holds such negative connotations in terms of the debate between Darwinism and creationism. In some countries, some people don't recognise the term Darwinism as referring to 'evolution' as a whole, but instead think that it refers only to human evolution, and others perceive Darwinism to be intrinsically atheistic.

Darwin was obviously an immensely important historical figure, but should we even expect all evolutionary biologists to have read *On the Origin of Species*? Though this is something I would wholeheartedly advocate, there are a number whom I suspect have not. But then we don't expect physicists or chemists to be fully conversant in the work of the historical figures of their field, or to call themselves Einsteinian or Lavoisierian. So one has to wonder, when the mechanisms described in the nineteenth-century theories have been expanded and developed for between 150 and 200 years, does it really mean anything any more to call someone a Darwinian, an anti-Darwinian or a Lamarckian? Surely the only purpose these labels can serve is to define an antithesis to one's own perspective. Over the past 60 years they have really acted, in the public sphere at least, as a rather crude shorthand for who is 'in' and who is 'out' in the science community, or to denote those on the most radical side of cutting-edge – to some extent they are applied regardless of which parts of Darwin's actual theories are being upheld. Of course, the problem is that these terms have now taken on a life of their own in the Darwinism vs. creationism debate. And to be labelled anti-Darwinian has now, more than ever, taken on a whole new meaning – it now carries with it the stigma of pseudo-science or, at worst, of fuelling anti-scientific sentiment.

Anti-science – the threat of creationism

The wider worldview that Dawkins is advocating is, of course, evolutionary theory and science itself. On the Richard Dawkins Foundation website, Dawkins states that part of its mission is to defend science 'from deliberate attack from organized ignorance. We even have to go out on the attack ourselves, for the sake of reason and sanity.'

Presumably by 'organized ignorance' Dawkins is referring to creationism. Creationism as a political movement is apparently an increasing threat to Dawkins' worldview. While it is undeniable that this is an important subject that needs to be tackled, if you believe the hype that creationism is on the rise – not only is it taking over in the US but it is on the move to Europe and now, according to Dawkins *et al*, we have the threat of 'Islamic creationism' to combat. It may be the case that we are becoming more aware of 'creationist' viewpoints beyond the US. And while I don't doubt for a second that this is a cause for concern, one has to ask how much of this is due to a rise in creationism and how much to a rise in media attention – after all, creationism is not always a majority viewpoint or even one that is a clearly defined discourse in some countries. It may not even be that the problems we are facing are solely about organised religious reactions to evolutionary theory, but relate more closely to issues of wider public awareness and declining science education. If we focus merely on the negative of 'creationism', it creates a dichotomy that will inevitably lead to conflict and hyperbole. In a way it might actually be creating the problem of rising creationism. Unfortunately, this can also mean we overlook the valuable work that is being done by many people to redress the balance by effectively raising the public understanding of evolutionary theory with new audiences.

As was highlighted in chapter 6 there simply has not been enough large-scale longitudinal research carried out into the impact of creationism and the public understanding of evolutionary theory on a global scale – though more is clearly needed. Most of the coherent research into the effect of creationism on the acceptance of evolutionary theory has understandably taken place in the US. Some recent studies there have suggested that far from being on the rise, 'creationism' has maintained a stable rate or may in fact be in decline. A 2007 study undertaken by the Coalition of Scientific Societies on 'creationism' in the US showed some interesting results. The summary of the research was that 'most respondents accepted that life evolved, many accepted that it evolved through natural processes, and more favored teaching evolution than creationism or intelligent design in science classes'. While a significant and worrying number of respondents still favoured creationism, these results (among others) would seem to suggest that this was not a majority viewpoint. Interestingly, they also showed that a significant number of respondents thought that evolution had occurred under guidance from a deity. Concern over the threat of 'creationism' has often led to the depiction of creationists holding a growing majority in the US, while scientists are under siege, though this may be an understandable by-product of some aspects of the Bush administration's approach to science policy.

Obviously this is a debate in which Dawkins has been very publicly involved. In a 2008 article in the *Guardian*, when Decca Aitkenhead asked him 'if he feels that public understanding of science has improved during his career', he responded: 'I would say that when my academic career began there was probably just as much ignorance – but less active opposition [to science].' He then went on to suggest that there is a 'a fairly substantial fraction of young people' who reject evolution without thinking due to the way in which they have been brought up. This, Dawkins says, is 'due to greater religious influence' rather than problems with scientific education.

It would indeed be cause for concern if students were turned off from evolutionary theory because they came from religious backgrounds, but if we continue to inappropriately promote evolutionary theory as an 'atheistic' challenge to religion, what else can we expect?

The God Delusion openly seeks to convert readers to atheism. It is therefore paramount for Dawkins to show that there is a need to convert people, particularly when it comes to his area of expertise – evolutionary theory – regardless of whether there has actually been a significant rise in 'creationism' or whether this promotion of evolution as part of a wider atheist movement might actually

THE SWORD OF RHETORIC

contribute to the problem. But we should ask ourselves: is this effective science communication, or proselytising?

The Age of Dawkins

The history of evolutionary theory is full of examples of the popular advocacy of theories that may or may not have been widely accepted by the scientific mainstream. Likewise, history books are littered with the opinions of a multitude of scientists whose work we no longer remember today, regardless of how influential they were in their day. Those who stand the test of time tend to be those who engender or contribute to a continuing research programme or school of thought. Trends in popular science can be just as fickle as they are elsewhere. In order to communicate science effectively, it is necessary to adapt the content to take account of current research. To be fair to Dawkins, he probably did not think when he was writing books like *The Selfish Gene* and *The Extended Phenotype* that they would endure for so long. This, however, throws up a problem for the way in which Dawkins has used advocacy in some of his works – the advocating of a position implies a certain inflexibility. A position that one advocates merely in order to inspire is not something that can easily be countered by research data, as it makes no real predictions or claims for research outcomes. Conversely, it is potentially difficult for a concept which is merely advocated to encompass new research challenges in a meaningful way. This way of setting up a scientific concept differs in many ways from how most scientists think the practice of 'science' should ideally be undertaken or debated. Should we force the data to fit the metaphor, or develop new metaphors to reflect the research data? Flexibility in response to changing debate and research cannot coherently be achieved by those who vehemently advocate only one position, which is not in any way testable. If any scientist were to stake his or her entire career on advocating a worldview, a metaphor or a concept, it undoubtedly runs the risk of becoming outdated. Their concepts may well have to be continually stretched to breaking point to account for new research challenges. On the other hand, scientists who seek to communicate and contribute to the current state of play in their discipline or field, as most hope to, are not wedded to concepts that will age as they do.

Advocacy as a science communication tool is really a top-down, paternalist approach – it is simply not interactive enough and would not be the tool of choice for many in either scientific publishing or popularising. It can also imply a role for a kind of certainty about theories or concepts that goes beyond the realm of science. Concepts such as 'certainty', 'truth', 'theory' and 'fact' need to be discussed with more transparency and clarity by scientists when explaining their work. In an ideal world, science communication would be analogous to teaching and would therefore seek to do everything a good teacher does for their pupils – to equip the audience with the critical and analytical tools, and the information, to make their own informed, well-reasoned and educated decisions about which stance they wish to adopt. And as any good teacher or lecturer knows, teaching is a two-way process. In light of this, real attempts should be made to engage with people's concerns or criticisms – they should not be simply derided. There is a heavy responsibility, incumbent on both the science community and those who communicate scientific research, to take joint responsibility for the way in which their disciplines are promoted. Advocacy and rhetoric can all too easily turn into proselytising. With no adequate way to discern between the differing perspectives that can be circulated by members of the scientific community, is it surprising that members of the public can find it hard to work out who to trust?

Another aspect of the proselytising effect of advocacy is the need for serious debate about whether science should be promoted as atheism to the extent that Dawkins does. By labelling whole swathes of the population as anti-Darwinian, anti-enlightenment or anti-reason, what is Dawkins actually achieving? The fact that this approach is potentially counter-productive can no longer be ignored; it surely serves only to entrench division rather than to move the debate forward. It is not the purpose of any book to tell the reader what to think, but as we begin to face up globally to some of the most serious challenges of our generation – among them loss of biodiversity, climate change and international terrorism – should we not be seeking reconciliation, and attempting to recognise our shared agenda in the face of issues that will affect us all? To do this, we need to positively highlight the pivotal role that science and the understanding of science plays in society. Those on either side of the faith debate need to work together to dispel misinformation about science, and to challenge detrimental superstitions and misconceptions.

CONCLUSION

LOOKING FORWARD

Ultimately, I agree with Dawkins on many things – most fundamentally, the importance of raising awareness and understanding of evolutionary science. Where we part company is on how best that can be achieved. Dawkins has now retired from his post as the Charles Simonyi Professor for the Public Understanding of Science at Oxford University and been replaced by the mathematician Professor Marcus du Sautoy. So what tips should Richard Dawkins' replacement bear in mind?

History of science – context is key

First, he should be careful about how he portrays the history of his discipline and the practice of science. The debates surrounding science and religion and the history of evolutionary theory are much more complex than they would appear to be from Dawkins' depiction of them. While my purpose here has not been to downplay the serious challenges that Darwinism posed for theological and scientific thinkers alike in the nineteenth century, it is not possible to remove the 'science' of this period from its ideological context. As we have explored, there is much more to the history of evolutionary biology than a shift from a biblical version of creationism to a Darwinian worldview. Rather than there being a simple dichotomy between 'evolution' and 'religion', there were a number of successful attempts to reconcile these two perspectives. Various other debates had an impact on the way in which evolutionism was perceived and subsequently promoted, not least the professionalisation of science and the increased specialisation into separate disciplines that was a key factor in the late nineteenth century and the early twentieth century.

There existed a number of ideas surrounding the non-fixity of species before Darwin, and he was not the first to suppose that species could change over time. Nor was he the first or only person to put forward theories of natural selection. Contrary to what you might expect, there are some significant differences between Dawkins' and Darwin's versions of evolution. The most profound of these relate to inheritance and the source of variation. Darwin accepted not only use inheritance, but also the effect of the environment on characteristics which are passed down the generations. To some, the term 'Darwinism' might cover a much broader definition of evolution – say, a theory of common ancestry or descent with modification – which does not necessarily imply a sole role for natural selection and a heavy focus on an adaptationist model. I suspect that to Dawkins it is framed much more in terms of adaptation and natural selection than this. One should be careful as to how the term 'Darwinism' is used, as it has throughout the years had many different meanings.

The overriding atmosphere of progress that pervaded the profound social, political and technological shifts in Darwin's day filtered through into evolutionist thinking. It is this focus on a progressional model of evolution that is out of synch with modern evolutionary thought, rather than any historical backlash from theologians. Dawkins is fighting two battles rather than one. He also seeks to expunge any evolutionary ideas that are tainted by historical association with the concept of a progressional or goal-directed evolution from our perspective of Darwinism and evolutionary biology as a whole. Might this be throwing the proverbial baby out with the bathwater? The threads of non-neo-Darwinian thought have continued to run throughout the twentieth century into the twenty-first. There are those today who challenge not necessarily natural selection per se, but the idea that natural selection is the sole significant mechanism for evolutionary change. The recent focus on gene-centric thinking has perhaps to some extent stifled research in other directions. History – and particularly types of history that are skewed towards the dominant contemporary research perspective – can act to constrain thought and the communication of key ideas. Dawkins must surely know this as much as anyone else, and the aim of giving further credence to his neo-Darwinian perspective must surely have crossed his mind when he made the selections for his 2008 anthology *The Oxford Book of Modern Science Writing.*

The late-nineteenth-century preoccupation with a progressional or goal-directed model of evolution is long gone, but some of the important questions that are sometimes erroneously and negatively associated with this school – specifically that of neo-Lamarckism – are still with us today. The most recent of these are the products of debates on horizontal gene transfer and evolutionary developmental biology, and of the re-evaluating of the role of the environment in evolutionary mechanisms and the models of heredity that are central to Dawkins' version of classical modern genetics. And while these may in time be seamlessly incorporated into an expanded Darwinian perspective, the current resurgence in research in these areas should teach us one thing – that we should approach new developments in science with an open mind.

Be a PEST

Science communication is a very valuable resource, not only for society but also for scientists – and this is becoming much more widely recognised. In this age of web-based information, it should perhaps move away from a top-down paternalistic approach and aim to provide the tools for the public to critically engage with the sciences on their own terms. Science communication isn't simply about presenting one's own hypotheses or worldview. Regardless of the historical perspectives on how Dawkins' work relates to the representation of 'Darwinism', there are still currently a number of scientific challenges to Dawkins' selfish gene hypothesis which seem to have been neglected by popular science accounts of evolutionary theory. If this is the case, there are a number of reasons for it, including the 'threat' of creationism and the grip of 'brand Dawkins' on the public perception of evolution. Communication is an important factor not only in the way in which the public engages with the scientific community, but also in the way in which different scientific disciplines engage with each other. Science is not an amorphous mass entity. Science is a group activity involving a number of different communities. One issue that becomes self-evident when exploring the history of a concept like 'evolution' is that different disciplines understand the concept in different ways depending on their research focus. A microbiologist will look at heredity or even the concept of what a species *is* in a very different way to a zoologist. Beyond the biological sciences, a computer scientist might understand a selective process in a very different way to a physicist.

In addition to this, with the current concerns over public loss of trust in 'science' a top-down approach to communicating science is probably not the best, and certainly not the only, method of communication that should be employed. Communicating science is a two-way process. It is no longer the case that the audience can be seen as an uninformed mass waiting to be converted to a scientific perspective. People have genuine and valid concerns about the impact of and implications for science in society. If scientists tell people what to think and do not allow any engagement, then they will ultimately end up with a lesser audience of core followers and a sense of disenfranchisement in the rest. Preaching to the converted is narrow-minded, and serves only to entrench divisions rather than effectively challenging them.

Critical and analytical awareness is as much at issue as scientific literacy. Using the term 'truth' in the context of science or implying certain and unshakeable knowledge comes with its own pitfalls. We are all so used to stories in tabloid newspapers telling us one week that something causes cancer and the next that it fights cancer, that it is understandably difficult to unpick what is worth believing. This can have catastrophic effects – as with the MMR scandal. How can the public easily decide what is 'bad' science and what is 'good' science if we don't honestly lift the veil from how different scientists and those who study the practice of science think that it really works? This is where the PEST (public engagement in science and technology) model of science communication can come into its own, for it allows for a much more accessible and engaging forum for debates about not only the content of science but also its practice. If we should be wary of using the term 'truth' in relation to science, then to appropriate it for discourses that go beyond the realm of science leads to further problems still.

The humanities can play a part in communicating the sciences, by playing a role that is complementary to the sciences. If we are to have a truly enlightened society we need to engage as much with the context of science as with the practice of science. This is contrary to Dawkins' 'only light in the darkness' version of science, which is very much born of an outdated and overly simplistic 'science wars' perspective. The unfortunate rejection of a humanities-based approach by some science communicators – due to a misplaced apprehension that the world of the humanities is, broadly speaking, a hotbed of arch-relativists – has perhaps been a little counter-productive. Fortunately, with the current rise in coherent multidisciplinary research and funding, this conflict between the sciences and humanities, and the lack of trust it has engendered, is hopefully receding. Perhaps we are seeing the beginnings of a paradigm shift? Let's hope so – because, worryingly, an increasing number of scientific commentators are expressing the idea that the real organised creationist threat comes from Islam. This is to some extent an anecdotal perspective that is as yet unsubstantiated by sufficient research. It is a stance that is dangerously close to being a political one and is not one that is always born of substantial or even indicative sociological research.

In certain areas there can be a clear role for those outside the research sciences to be involved in debates about the role of science – an obvious one is medical ethics – and these are valid discussions that are already widely recognised by most scientists. For example, the ethics curriculum at various medical schools in the UK has been enlarged in response to a recognised need for ethics to be part of medical training. The waxing and waning social mistrust of those claims of 'science' that are thrown around in the media – exemplified by the MMR scandal – and the portrayal of 'science' as part of a wider 'clash of cultures' by Dawkins and others, clearly show that science communication has a very important political role to play in society. To address this, the whole scientific and academic community needs to engage coherently with the public on all areas of research. There also needs to be a wider recognition of the public's role in these important debates.

While there is a place for books that give an inside perspective on new ideas, research directions or themes – as *The Selfish Gene* and *The Extended Phenotype* did – it becomes slightly problematic when the important role of science communicator is used to promote a highly personalised campaign. There has been a slow process in Dawkins' writing that led us from *The Blind Watchmaker*, through *Unweaving the Rainbow* and *A Devil's Chaplain*, to the almost inevitable *The God Delusion*. And unfortunately, after a lot of interesting, accessible and thought-provoking work, Dawkins lost some of us at *The God Delusion*.

Mind your language

Another matter that the new Charles Simonyi professor should consider carefully is the way in which he uses language in his writing. The language that is used to communicate often very subtle differences in scientific ideas can in its own way act to constrain not only the public understanding of key concepts, but also the direction of research in other fields. After all, as Dawkins himself has modestly pointed out, we cannot all be expected to read every book in every discipline that relates to the subject of evolution. So even those who practise the field of evolutionary biology may have different perspectives on certain aspects of their wider discipline – this is only to be expected. It is therefore doubly important that all aspects of a discipline are communicated – lest those first entering a discipline become entrenched in one perspective over and above another equally valid one. Remember, popular science books are read by scientists as well. A good science communicator can act as a 'translator' between disciplines and fields as well as ably engaging the wider public.

Language can also play another role in science communication; for example, the labels used to describe competing theories or researchers can act as constraints on them. Woe betide anyone who was labelled a Lamarckian in the late twentieth century

– this could signal the end of your career. In the past few years this label has become less of a slur in certain academic circles due to increased interest in emergent areas of non-gene-centred research; however, it still carries a significant stigma among the public and some scientific circles. Dawkins has used this term to imply that it is a heresy – but surely there can only be heresy where there is dogma? Darwinism is also used as a label – but one that denotes acceptability and conformity. Yet aspects of Darwin's theories would be unrecognisable to some neo-Darwinians today. Just as Darwin was no Darwinian, Weismann was no Weismannian and I am sure there are plenty who would argue that Mendel was no Mendelian! Dawkins has rebranded Darwinism as an atheistic stance, but on this and in other areas I think he and Darwin would strongly disagree.

The labels that we apply to research can be used as slurs as well as to communicate ideas effectively. Darwinism and Lamarckism work very much in this way. To label something 'Darwinian' is to give it a seal of approval, and to label something as 'Lamarckist', group selectionist or culturally relativist is to invoke a shorthand for dubious science or shoddy, politically motivated logic. This has been a common theme, from Dawkins labelling new research as 'a Lamarckian scare' in *The Extended Phenotype* to more recent concerns over Wallace the arch-selectionist being labelled a group selectionist with regard to his 1858 paper. Conversely, while it is tempting for the popular press to label things as anti-Darwinian, this can sometimes give easy ammunition to those who wish to maintain the status quo. Examples of this could be seen in the 1980s and 1990s, in a number of 'Lamarck versus Darwin'-style articles reporting various groups' research in national newspapers, right through to the recent *New Scientist* 'Darwin was Wrong' shocker. However, I seriously doubt that this kind of reporting will really make much of a difference in the increasingly political and complex 'creationism' vs. 'Darwinism' debate. Even if this were the case, it is debatable whether we should really avoid discussing these challenges publicly in a way that seeks to interest people in new research, as long as the reporting is fair and balanced. It is much easier to take umbrage at this style of journalism than to publicly engage with the challenges the stories embody.

The new professor for the public understanding of science should also be more aware of cultural differences when it comes to communicating science or understanding its context, be that historically, politically or sociologically. Science may be objective in its relationship with its subject matter, natural phenomena, but it cannot be removed from its context. As we have discussed, Darwinism means many different things to many different people, to the point where it has become a problematic term for communicating evolutionary science. In some ways it is so heavily associated with an atheist agenda that it can close down dialogue and engagement. The cross-cultural context of language can lead to all sorts of problems, as researchers at McGill University in Canada discovered when investigating perspectives on biological evolution in Indonesian and Pakistani communities. They found that the term 'ancestor' can be translated into their native languages in a number of ways which can give the term very different meanings, ranging from immediate ancestors – like grandmothers – to spiritual ancestors or historically defined religious ancestors. This of course has the potential to lead to confusion over concepts that, in Europe, we might see as being easy to understand. The subject matter of science may not be culturally or historically relative, but that language with which we communicate it most certainly is.

Perhaps the most important point when it comes to language is that there is really no excuse for insulting or intemperate language when it comes to dealing with one's critics. If we ridicule them or make personally offensive comments, how can we truly expect people to engage with our point of view? Coupled with this, the employment of rhetoric that seeks to polarise rather than unpack complex issues can be very divisive and ultimately counter-productive. This is apparent in all spheres of communication, whether it is terms like 'enemies of reason' or 'axis of evil'. When you work in the public sphere in the way that Dawkins has done increasingly throughout this career, this carries with it the same responsibility of respectful communication that we might also wish to hear from our politicians.

Science is not atheism

Science vs. religion is a false dichotomy, which does not truly reflect the rich history of the sciences and increases the polarisation between different approaches to understanding the natural world. Evolutionary theory and other areas of research, like physics, mathematics or chemistry, cannot and should not claim to disprove people's faith in a divinity. Evolutionary theory does disprove concepts like young earth creationism, but it cannot disprove 'God' per se. Dawkins is not anti-religious because acceptance of evolutionary theory precludes all religious belief – he is an atheist because he chooses to be one. Can we truly claim to live in a free society if we are not free to choose what we believe? I am sure that both Dawkins and I would respond 'no' to this. However, I doubt that he would go as far as I do in saying that this cuts both ways and we should respect other people's faith. It is counter-productive to paint evolutionary theory as an atheistic stance – this closes down dialogue and engagement with faith groups that have historically not had an ongoing problem with accepting evolutionary theory. Furthermore, one should be careful to discriminate between a philosophical discourse and a scientific reasoning. It is a different proposition to argue from a philosophical perspective that there is probably no God (and here the term *probably* is most important), but one cannot assert this as a scientific fact or even a scientific hypothesis.

There are obviously some fairly difficult discussions to be had, specifically in relation to evolution and science education. But while such matters clearly have religious implications, it is too simplistic to paint this as a straight 'religion vs. science' debate. In 2005 a trial in Dover, Pennsylvania in the US that focused on whether a school board was wrong to decide to teach intelligent design in biology classrooms was successfully brought about by eleven parents suing the school. Included in this admirable group of plaintiffs were parents who were themselves from religious backgrounds, but who did not wish to see religious perspectives taught in a science classroom.

While it might be enticing to see evolutionary theory as the beginning of an as yet unfulfilled enlightenment drive towards the 'truth' of atheism, this is not the case. This claim fundamentally undermines both the nature of the enlightenment ethos and the many thinkers who have worked or do work within the sciences and humanities, who either see themselves as agnostic or adhere to a religious or spiritual faith.

Science and society – the way forward

In conclusion, there are currently a number of potential scientific challenges to Dawkins' selfish gene hypothesis which have possibly been neglected by popular science accounts, and as such have not penetrated the public consciousness to the same level as Dawkins' neo-Darwinian perspective. Hopefully, if it has achieved anything this book has shown that you don't have to be a creationist to be anti-Dawkins, you don't have to be anti-science to be anti-Dawkins and you most certainly don't have to be anti-Darwin to be anti-Dawkins. Even before this book has been finished I have been criticised on web forums for writing a critical book about Dawkins – it is instantly assumed, without any knowledge of the content of this book, that I must be writing from a theistic position or an anti-evolution stance. It was even suggested by one over-zealous wag that I had in some way fraudulently obtained my PhD or failed in my undergraduate studies. I don't expect that I will persuade many of Dawkins' band of strong supporters to reconsider, and nor am I seeking to – they are entitled to their opinion as much as anyone else. Hopefully, if nothing else I have succeeded in giving a voice to the many people who happily support evolutionary theory, but increasingly find Dawkins more a counter-productive hindrance to debate than an a useful ally.

So far 2009 has seen an unprecedented interest in all things related to Darwin and evolution. And not only in Darwin's native Britain – the global response to the anniversary celebrations has been encouraging and overwhelming. Being involved in these celebrations has led me personally to challenge a lot of the assumptions I had inadvertently made about the reception of Darwinism internationally. I was at the American Association for the Advancement of Science's annual conference in Chicago on 12 February 2009 – the bicentenary of Darwin's birth. While I was there I was fortunate enough to be invited to an Episcopalian church in Chicago to a rather surprising Darwin birthday celebration. This birthday party – there was even a cake – was part of the 'Evolution Weekend Clergy Letter' project. Started by Michael Zimmerman, a professor of biology at Butler University in Wisconsin, this project has been supporting celebrations of Darwin's birthday since 2006, and also seeking signatures from the American clergy for a letter in support of evolutionary theory. The letter states that 'We believe that the theory of evolution is a foundational scientific truth, one that has stood up to rigorous scrutiny and upon which much of human knowledge and achievement rests. To reject this truth or to treat it as "one theory among others" is to deliberately embrace scientific ignorance and transmit such ignorance to our children.' As this book goes to press, Zimmerman's group has collected nearly 12,000 signatures from American Christian clergy.

Hopefully the age of empty 'clash of cultures' rhetoric is coming to an end. Perhaps the scientific community, and especially its most high-profile ambassadors, would do well to emulate the approach of US President Barack Obama to working towards solutions to huge global problems in a sometimes polarised intellectual climate. Speaking at Notre Dame University in May 2009, he made some points which I believe could usefully be applied to the debates about science, faith and society, especially in these times of crisis:

> We must find a way to reconcile our ever-shrinking world with its ever-growing diversity – diversity of thought, of culture, and of belief.
>
> [...] For the major threats we face in the twenty-first century – whether it's global recession or violent extremism; the spread of nuclear weapons or pandemic disease – do not discriminate [...] Our very survival has never required greater cooperation and understanding among all people from all places than at this moment in history. Unfortunately, finding that common ground [...] is not easy.

[...] Each side will continue to make its case to the public with passion and conviction. But surely we can do so without reducing those with differing views to caricature. Open hearts. Open minds. Fair-minded words.

SELECTED NOTES, REFERENCES AND FURTHER READING

Introduction

See Richard Dawkins, *The God Delusion* (Bantam Press, 2006); Richard Dawkins' description of survival of the fittest in 'Chapter 5: an agony of five fits', *The Extended Phenotype: The Long Reach of the Gene* (Oxford Paperbacks, 1999), second edition; quote on Fisher found in the preface to Richard Dawkins, *The Selfish Gene* (Oxford Paperbacks, 1989), second edition, p. xi. For a comprehensive account of Gould on the hardening of the modern synthesis see Stephen Jay Gould, 'Chapter 7: the modern synthesis as a limited consensus' in *The Structure of Evolutionary Theory* (Harvard University Press, 2002); for an anniversary account of the influence of the publication of *The Selfish Gene* which includes a chapter by Daniel Dennett entitled '*The Selfish Gene* as philosophical essay' see Alan Grafen and Mark Ridley, *Richard Dawkins: How a Scientist Changed the Way We Think: Reflections by Scientists, Writers, and Philosophers* (Oxford University Press, 2006); Steve Jones, *Darwin's Island: The Galapagos in the Garden of England* (Little, Brown, 2009); Stephen Jay Gould, 'Darwinian Fundamentalism', *The New York Review of Books* (12 June 1997), vol. 44, no. 10.

For comments on Lamarckism and junk DNA see Richard Dawkins, 'Chapter 9: Selfish DNA, Jumping Genes and a Lamarckian Scare'; for comments on advocacy see 'Chapter 1: Necker Cubes and Buffaloes' p. 1; for central theorem see 'Chapter 13: Action at a Distance', p. 233; for Dawkins on Steven Rose and genetic determinism see 'Chapter 2: Genetic Determinism and Gene Selectionism', pp. 10 and 28, all in *The Extended Phenotype: The Long Reach of the Gene* (Oxford Paperbacks, 1999), second revised edition. See also Richard Dawkins, *Climbing Mount Improbable* (W.W. Norton, 1996); William Paley, *Natural Theology: or, Evidences of the Existence and Attributes of the Deity* (1802); Richard Dawkins on Paley in *The Blind Watchmaker* (Penguin Books, 1991), pp. 5–6; Charles Percy Snow, *The Two Cultures* (Cambridge University Press, 1959); Richard Dawkins, *River out of Eden: A Darwinian View of Life* (Weidenfeld and Nicolson, 1995); see also Marek Kohn, *A Reason for Everything: Natural Selection and the English Imagination* (Faber and Faber, 2005), new edition.

Chapter 1

Erasmus Darwin, *The Temple of Nature, or, The Origin of Society* (J. Johnson, published posthumously, 1803); Aristotle on Empedocles in Francis Cornford, *Plato's Cosmology* (Kegan Paul, 1937); al-Jahiz quote in Ehsan Masood, *Science and Islam: A History* (Icon Books, 2009). For a good account of early Greek science see Andrew Gregory, *Eureka! The Birth of Science* (Icon Books, 2001); Ulisse Aldrovandi quoted in Paula Findlen, *Possessing Nature* (University of California Press, 1996), p. 248. For further reading on natural history in the Renaissance see Brian Ogilvie, *The Science of Describing: Natural History in Renaissance Europe* (Chicago University Press, 2006); for further reading on the Scientific Revolution see Lisa Jardine, *Ingenious Pursuits: Building the Scientific Revolution* (Abacus, 2000); the seminal challenge to the concept of a 'scientific revolution' can be found in Steven Shapin, *The Scientific Revolution* (Chicago University Press, 1998). On Hans Sloane's collection see the British Museum website: http://www.britishmuseum.org/the_museum/history_and_the_building/sir_hans_sloane.aspx

See also Toby Musgrave, Chris Gardner and Will Musgrave, *The Plant Hunters: Two Hundred Years of Adventure and Discovery* (Ward Lock, 1999); of less relevance but equally good is Toby and Will Musgrave, *An Empire of Plants: People and Plants that Changed the World* (Cassell and Co, 2000); for Lyell see Charles Lyell, *Principles of Geology* (first edition, 1830–33, and tenth edition, 1867–8); for further reading see Derek Blundell and Andrew Scott (eds), *Lyell: The Past is the Key to the Present* (Geological Society Publishing House, 1998); for a comprehensive account of naturalists and geologists in this period see Peter Bowler, *Evolution: the History of an Idea* (University of California Press, 2003), third revised edition.

On Francis Bacon see Paolo Rossi (trans. Sacha Rabinovitch), *Francis Bacon: from Magic to Science* (Routledge & K. Paul, 1968); 'Cogito ergo sum' in René Descartes, *Principia philosophiae (1644)*; for a good account of the debate between 'rationalists' and 'empiricists' see Peter Markie, 'Rationalism vs. Empiricism' (2008) and the online *Stanford Encyclopedia of Philosophy* at http://plato.stanford.edu/entries/rationalism-empiricism/

For a detailed analysis of the reception of Paley's work see Aileen Fyfe, 'The reception of William Paley's *Natural Theology* in the University of Cambridge', *British Journal for the History of Science* (1997), vol. 30, pp. 321–35. For a good account of the historical context of debates in science and religion see John Hedley Brooke, *Science and Religion: some historical perspectives*

(Cambridge University Press, 1991). See also Adam Smith, *An Enquiry into the Nature and Causes of the Wealth of Nations* (1776).

A more general account of the history of science can be found in John Gribbin, *Science and History* (Allen Lane, 2002) and Patricia Fara, *Science: A Four Thousand Year History* (Oxford University Press, 2009). For a classic account of the development of evolutionary thought see Henry Fairfield Osborn, *From the Greeks to Darwin: an Outline of the Development of the Evolution Idea* (Macmillan, 1894).

Chapter 2

Charles Darwin, *The Descent of Man, and Selection in relation to sex* (John Murray, 1871), first edition, quote taken from p. 3; Thomas Malthus, *An Essay on the Principle of Population; or, a view of its past and present effects on human happiness; with an inquiry into our prospects respecting the future removal or mitigation of the evils which it occasions* (1826); Charles Darwin in Francis Darwin (ed.), *The Life and Letters of Charles Darwin, Including an Autobiographical Chapter* (D. Appleton and Company, 1887), vol. I, quote taken from p. 87; Ernst Mayr, *The Growth of Biological Thought* (Harvard University Press, 1982), quote taken from p. 330. On Buffon see Jacques Roger (trans. Sarah Lucille Bonnefoi), *Buffon: A Life in Natural History* (Cornell University Press, 1997); Philip Sloan, 'The Buffon-Linnaeus Controversy', *Isis* (1976), vol. 67, pp. 356–375. For sections on Buffon, Cuvier and Saint-Hilaire see Peter Bowler, *Evolution: the History of an Idea* (University of California Press, 2003), third revised edition; quote on p. 50 from Charles Darwin, *On the Origin of Species by means of natural selection, or the preservation of favoured races in the struggle for life* (John Murray, 1866), fourth edition, p. xiii; quote on pp. 52–3 from Charles Darwin, *On the Origin of Species by means of natural selection, or the preservation of favoured races in the struggle for life* (John Murray, 1859), first edition, pg. 434. Further reading: Ralph O'Connor, *The Earth on Show: Fossils and the poetics of popular science 1802–1856* (Chicago University Press, 2008).

See Jean-Baptiste Lamarck (trans. Hugh Elliot), *Philosophie Zoologique* (Macmillan, 1914); Peitro Corsi, *The Age of Lamarck: Evolutionary Theories in France 1790–1830* (University of California Press, 1988), p. 11. For Patrick Matthew quote see Patrick Matthew, *Naval Timber and Arboriculture* (Longman, 1831); Matthew's letter to the *Gardeners' Chronicle*, 12 May 1860 is cited in Kentwood D. Wells, 'The Historical Context of Natural Selection:

The Case of Patrick Matthews', *Journal of the History of Biology* (1973), vol. 6, no. 2, pp. 225–258; the letter from Darwin to the *Gardeners' Chronicle* is in Francis Darwin (ed.), *The Life and Letters of Charles Darwin, Including an Autobiographical Chapter* (1887), vol. II, p. 301; quote on p. 63 from Charles Darwin, *On the Origin of Species by means of natural selection, or the preservation of favoured races in the struggle for life* (John Murray, 1861), third edition, p. xv.

See Jim Secord, *Victorian Sensation: the Extraordinary Publication, Reception, and Secret Authorship of 'Vestiges of the Natural History of Creation'* (University of Chicago Press, 2000); quote on p. 66 from letter from Darwin to Hooker in Francis Darwin (ed.), *The Life and Letters of Charles Darwin, Including an Autobiographical Chapter* (1887), vol. I, p. 333; Disraeli quote on p. 65 from Benjamin Disraeli, *Tancred* (1847), p. 93; quote on p. 67 from Charles Darwin, 'An Historical Sketch of the recent progress of opinion on the Origin of Species' in *On the Origin of Species by means of natural selection, or the preservation of favoured races in the struggle for life* (John Murray, 1861), third edition, p. 16.

See Alfred Russel Wallace, *My Life: A Record of Events and Opinions* (Chapman and Hall, 1905), vols. I and II; quotes on p. 68 from letters from Darwin to Lyell in Francis Darwin (ed.), *The Life and Letters of Charles Darwin, Including an Autobiographical Chapter* (1887), vol. II, pp. 116, 117; Alfred Russel Wallace, 'On the Law which has Regulated the Introduction of New Species', *Annals of Natural History* (1855), vol. 16, no. 2, pp. 184 to end; Charles Darwin and Alfred Russel Wallace, 'On the Tendency of Varieties to Depart from the Original Type and on the perpetuation of varieties and species by natural means of selection', *Journal of the Proceedings of the Linnaean Society, Zoology*, vol. III (20 August 1858); quote on p. 69 taken from Alfred Russel Wallace, 'Wallace's remarks on receiving the first Darwin-Wallace Medal on 1 July 1908', in *The Darwin-Wallace Celebration Held on Thursday, 1st July 1908* (1909). See also H. Lewis McKinney, *Wallace and Natural Selection* (Yale University Press, 1972); H. Lewis McKinney, 'Wallace's Earliest Observations on Evolution', *Isis* (1969), vol. 60, no. 3, pp. 370–373; quote on p. 70 from letter from Darwin to Wallace in Francis Darwin (ed.), *The Life and Letters of Charles Darwin, Including an Autobiographical Chapter* (John Murray, 1887), vol. II, p. 108; Herbert Spencer, *The Principles of Psychology* (1855); see also http://darwin-online.org.uk; http://darwinproject.ac.uk

For general background to Darwin and his work see Janet Browne, *Darwin's 'Origin of Species': A Biography* (Atlantic Books, 2007), new edition; and *Charles Darwin: Vol. 1 – Voyaging* and *Vol. 2 – The Power of Place* (Alfred A. Knopf, 2002); Adrian Desmond and James Moore, *Darwin* (Penguin, 1992), new edition.

Chapter 3

Thomas Henry Huxley, *Darwiniana: Essays* (D. Appleton, 1893), Westminster edition; Thomas Henry Huxley, 'Darwin on the origin of species', *The Times*, 26 December 1859, pp. 8–9; Herbert Butterfield, *The Whig Interpretation of History* (Hodder Arnold, 1931); quotes from Helena Cronin, *The Ant and the Peacock: altruism and sexual selection from Darwin to today* (Cambridge University Press) reprint edition, pp. 5 and 52; Samuel Butler, *Evolution Old and New: or The theories of Buffon, Dr Erasmus Darwin, and Lamarck as compared with that of Charles Darwin* (John Murray, 1882), second edition; Butler's influence is evident in George Bernard Shaw, *Back to Methuselah: A Metabiological Pentateuch* (Constable, 1925).

See Janet Browne, 'Darwin in Caricature: A Study in the Popularisation and Dissemination of Evolution', *Proceedings of the American Philosophical Society* (December 2001), vol. 145, no. 4, p. 82; quote from Charles Darwin, *On the Origin of Species by means of natural selection, or the preservation of favoured races in the struggle for life* (John Murray, 1859), first edition, p. 488; quote on pp. 78–9 from Thomas Henry Huxley, 'Darwin on the origin of species', *Westminster Review* (1860), vol. 17, pp. 541–70; J. Vernon Jensen, 'Return to the Wilberforce-Huxley debate', *British Journal for the History of Science* (1988), vol. 21, pp. 161–179; Frank James, 'An "Open Clash between Science and the Church?": Wilberforce, Huxley and Hooker on Darwin at the British Association, Oxford, 1860' in David Knight and Matthew Eddy (eds.), *Science and Beliefs: from natural philosophy to natural science, 1700–1900* (Ashgate, 2005); Charles Kingsley, *The Water Babies* (1863); Samuel Wilberforce, 'On the origin of species', *Quarterly Review* (1860), pp. 225–64; Thomas Henry Huxley, *Science and Hebrew Tradition: Essays* (Macmillan, 1893); Edward Fry, 'Darwinism and Theology', *Spectator* (1872), vol. 45, pp. 1137–38, 1168–1170 and 1201; Peter Bowler, *The Non-Darwinian Revolution: Reinterpreting a Historical Myth* (Johns Hopkins University Press, 1988); Peter Bowler, *The Eclipse of Darwinism: Anti-Darwinian Evolutionary Theories in the Decades around 1900* (Johns Hopkins University Press, 1983). See also Julian Huxley, *Evolution: The Modern Synthesis* (Harper and Brothers, 1942).

See Alvar Ellegard, *Darwin and the General Reader: The reception of Darwin's theory of evolution in the British periodical press 1859–1872* (Goteborg, 1958); George Saint Mivart, *On the Genesis of Species* (Macmillan, 1871); William Thomson, 'On Geological Time; of geological dynamics. Lectures read in 1868 to the Glasgow geological society', in *Popular Lectures and Addresses* (1868), vol II; Stephen Jay Gould, *The Structure of Evolutionary Theory* (Harvard University Press, 2002); H.C. Fleeming Jenkin, 'Review of *Origin of Species*', *The North British Review* (1867), vol. 46, pp. 277–318. For further reading see Peter Vorzimmer, 'Charles Darwin and Blending Inheritance', *Isis* (1963), vol. 54, pp. 371–390 and Peter Vorzimmer, *Charles Darwin: The Years of Controversy: The Origin of Species and its Critics 1859–1882* (Temple University Press, 1970).

For further reading see Michael Ruse, *Darwinism Defended: A Guide to Evolutionary Controversies* (Benjamin/Cummings Publishing Company, 1982), quote from p. 64; Richard Dawkins, *A Devil's Chaplain: Reflections on Hope, Lies, Science, and Love* (Houghton Mifflin Harcourt, 2003), pp. 67–9; Charles Darwin, *Variation of Plants and Animals under Domestication* (Murray, 1868), vols. I & II: quote on p. 91 taken from vol. 2, p. 357 (second edition, 1875); quote on p. 92 taken from vol. 2, p. 373. Peter Bowler, 'Darwin's Concepts of Variation', *Journal of the History of Medicine and Allied Sciences* (1974), vol. 29, pp. 196–212; Peter Bowler, 'Alfred Russel Wallace's Concepts of Variation', *Journal of the History of Medicine* (1976), vol. 31, pp. 17–29; Charles Darwin and Alfred Russel Wallace, 'On the Tendency of Varieties to Depart from the Original Type and on the perpetuation of varieties and species by natural means of selection', *Journal of the Proceedings of the Linnaean Society, Zoology* III, 20 August 1858; Edward Bagnall Poulton, *Charles Darwin and the Theory of Natural Selection* (Macmillan, 1901), quote from p. 80; Henry Fairfield Osborn, *From The Greeks to Darwin: An Outline of the Development of the Evolution Idea* (Macmillan, 1894); Charles Darwin in Francis Darwin (ed.), *The Life and Letters of Charles Darwin, Including an Autobiographical Chapter* (D. Appleton and Company, 1887), vol. I, quote taken from p. 93; George John Romanes, *Darwin and After Darwin: An Exposition of the Darwinian Theory and a Discussion of Post-Darwinian Questions* (Longmans, Green & Co, 1895), vols. I and II; Charles Darwin, 'Recollections of the Development of my Mind and Character', in

Francis Darwin (ed.), *The Life and Letters of Charles Darwin: Including an Autobiographical Chapter* (D. Appleton and Company, 1887), quote from vol. I, p. 93. For wider debates around the history of heredity see Staffan Müller-Wille and Hans-Jörg Rheinberger (eds.), *Heredity Produced: At the Crossroads of Biology, Politics and Culture 1500–1870* (The MIT Press, 2007). A full transcript of Dawkins speaking at the 2009 Open University annual lecture on 17 March 2009 can be found at http://www.open2.net/dawkins/

Chapter 4

See Frank Miele, 'Darwin's dangerous disciple: an interview with Richard Dawkins', *Skeptic* (1995), vol. 3, no. 4, pp. 80–85; Richard Dawkins, 'A Lamarckian Scare', chapter 9, *The Extended Phenotype: the Long Reach of the Gene* (Oxford Paperbacks, 1999), second revised edition. On Weismann see Rasmus Winther, 'August Weismann on Germ-Plasm Variation', *Journal of the History of Biology* (2001), vol. 34, no. 3, pp. 517–55; August Weismann (trans. and ed., with notes, by Raphael Meldola), *Studies in the theory of descent: with notes and additions by the author*, with a prefatory notice by Charles Darwin (Sampson, Low, Marston, Searle and Rivington, 1882); August Weismann, *The Germ-Plasm: A Theory of Heredity* (Walter Scott, 1893); August Weismann (trans. J. Arthur Thomson and Margaret R. Thomson), *The Evolution Theory* (1904), vols. I and II. In *The Extended Phenotype* Dawkins stated that 'the neo-Weismannist view this book advocates lays a stress on the genetic replicator as a fundamental unit of explanation' (Oxford Paperbacks, 1999), second revised edition, p. 113. See also Alfred Russel Wallace, *Darwinism: An Exposition of the Theory of Natural Selection with some of its Applications* (Macmillan and Co, 1889); George John Romanes, *Darwin and After Darwin: An Exposition of the Darwinian Theory and a Discussion of Post-Darwinian Questions* (Longmans, Green & Co, 1895), vol. II.

See Arthur Koestler, *The Case of The Midwife Toad* (Hutchinson, 1971); Paul Kammerer, 'Breeding Experiment on the Inheritance of Acquired Characters', *Nature*, 12 May 1923, vol. 111, no. 2793, pp. 637–40. Also see Joseph T. Cunningham, 'Breeding Experiments on the Inheritance of Acquired Characters', *Nature*, 26 May 1923, vol. 111, no. 2795, p. 702; Joseph T. Cunningham, *Hormones and Heredity: A Discussion of the Evolution of Adaptations and the Evolution of Species* (Macmillan, 1921); Nils Roll-Hansen, *The Lysenko Effect: The Politics of Science* (Prometheus, 2005); J.B.S. Haldane, 'Lysenko and Genetics', *Science and Society* (1940), vol. IV, no. 4, pp. 433–7.

See Francis Crick, 'On Protein Synthesis', *Symposia of the Society of Experimental Biology: The Biological Replication of Macromolecules* (1958), no. XII, pp. 138–63; for a challenge to Crick's dogma after Temin's work see Anon., 'Central dogma reversed', *Nature* (1970), vol. 226, p. 1198. For Crick's rearticulation of the central dogma see Francis Crick, 'Central Dogma of Molecular Biology', *Nature* (1970), vol. 227, pp. 561–63; for an overview of Steele's work see Edward Steele, Robyn Lindley and Robert Blanden, *Lamarck's Signature: How Retrogenes are Changing Darwin's Natural Selection Paradigm* (Allen & Unwin, 1998), new edition; for some of the earlier responses to Steele's work see Jonathan C. Howard, 'A Tropical Volute Shell and the Icarus Syndrome', *Nature* (1981), vol. 290, pp. 441–2; review by Laurence Hurst in *New Scientist*, 17 April 1999, pp. 60–61. For an excellent overview of the current work in epigenetics see Eva Jablonka and Marion Lamb, *Evolution in Four Dimensions: genetic, epigenetic, behavioral and symbolic variation in the history of life* (MIT Press, 2005), illustrated edition; quote from Eva Jablonka in *Taipei Times*, 11 February 2009, p. 9.

Chapter 5

The quote from Richard Dawkins, *The Blind Watchmaker* can be found on p. 155 of the 2000 reissue; Stephen Jay Gould, 'Evolution: The Pleasures of Pluralism', *New York Review of Books*, 26 June 1997, pp. 47–52; Richard Dawkins, chapter 9, 'Puncturing punctuationism', in *The Blind Watchmaker* (Penguin, 1991); Stephen Jay Gould, *The Structure of Evolutionary Theory* (Harvard University Press, 2002); Stephen Jay Gould, *Wonderful Life: The Burgess Shale and the Nature of History* (W.W. Norton, 1989). For a good account of the debates between Dawkins and Gould see Kim Sterelny, *Dawkins vs. Gould: Survival of the Fittest* (Icon Books, 2007), second revised edition.

W. Ford Doolittle, 'Phylogenetic Classification and the Universal Tree', *Science*, 25 June 1999, vol. 284, no. 5423, pp. 2124–8; Graham Lawton, 'Uprooting Darwin's Tree', *New Scientist*, 24 January 2009, pp. 34–9; Dawkins in post-lecture discussion for the Open University annual lecture, Tuesday, 17 March 2009 can be found at http://www.open2.net/dawkins/; for discussion on 'green beards' and kin selection see Richard Dawkins, *The Selfish Gene* (Oxford University Press, 1976) and Richard Dawkins, 'Twelve Misunderstandings of Kin Selection', *Zeitschrift für Tierpsychol* (1979), vol. 51, pp. 184–200; David Sloan Wilson and Edward O. Wilson,

'Evolution: Survival of the Selfless', *New Scientist*, 3 November 2007 and Danielle Fanelli, 'Altruism is no family matter', *New Scientist*, 12 January 2008. For Dawkins' response see Richard Dawkins, 'Comment: The group delusion', *New Scientist*, 12 January 2008 and at http://richarddawkins. net/article,2121,The-Group-Delusion,Richard-Dawkins

Also see Steve Connor, 'Evolutionists at war over altruism's origins', *Independent*, 10 January 2008; Darwin discussed cave-dwelling crabs and rats on pp. 137–8 in the section on 'Effects of Use and Disuse' in the first edition of *Origin of Species*. For further reading on Hox genes see Jason Scott Robert, *Embryology, Epigenesis, and Evolution: Taking Development Seriously* (Cambridge University Press, 2004); see also Micheal Syvanen and Clarence I. Kado, *Horizontal Gene Transfer* (Academic Press, 2002), second edition. Also see Lenny Moss, *What Genes Can't Do* (MIT Press, 2002), illustrated edition; Evelyn Fox Keller, *The Century of the Gene* (Harvard University Press, 2000); John Dupré and Barry Barnes, *Genomes and What to Make of Them* (Chicago University Press, 2008).

Chapter 6

Quote from Charles Darwin, *On the Origin of Species by means of natural selection, or the preservation of favoured races in the struggle for life* (John Murray, 1860), second edition, p. 481; letter from Charles Darwin to Asa Gray in vol. II, p. 311, quote from Charles Darwin's autobiography in vol. I, pp. 312–13, letter to Mr J. Fordyce quoted in vol. I, p. 304; quote concerning Dr Aveling's pamphlet in vol. I, p. 317; all in Francis Darwin (ed.), *The Life and Letters of Charles Darwin, Including an Autobiographical Chapter* (D. Appleton and Company, 1887). Madeleine Bunting, 'Darwin shouldn't be hijacked by New Atheists – he is an ethical inspiration', *Guardian*, 29 December 2008; Dawkins' response can be found at http://richarddawkins.net/articleCom ments,3475,Darwin-shouldnt-be-hijacked-by-New-Atheists---he-is-an-ethical-inspiration,Madeleine-Bunting-Guardian,page2; Mark Pallen, *The Rough Guide to Evolution* (Rough Guides, 2009) first Thus edition; Mark's blog can be found at roughguidetoevolution.blogspot.com/

Richard Dawkins, 'I'm an atheist, BUT…' can be found at http://rich-arddawkins.net/articleComments,318,Im-an-atheist-BUT---,Richard-Dawkins,page2; see also Richard Dawkins, preface to the paperback edition and discussion of Inca girl in chapter 9, *The God Delusion* (Bantam Press paperback edition, 2006); Francis Bacon, *The Advancement of Learning*

(1605); Richard Dawkins, 'Religion's misguided missiles', *Guardian*, 15 September 2001; Richard Dawkins, 'Bin Laden's victory', *Guardian*, 22 March 2003; Richard Dawkins, *The Root of All Evil?*, Channel 4, first screened January 2006; Manhattan skyline comment in the preface to *The God Delusion* (Bantam Press, 2006). For more information on female genital mutilation see S.D. Jones, J. Ehiri and E. Anyanwuc, 'Female genital mutilation in developing countries: an agenda for public health response', *European Journal of Obstetrics & Gynecology and Reproductive Biology* (2004), vol. 116, pp. 144–51; E. Sakeah, A. Beke, H.V. Doctor and A.V. Hodgson, 'Males' Preference for Circumcised Women in Northern Ghana', *African Journal of Reproductive Health* (2006), vol. 10, issue 2, pp. 37–47; Richard Dawkins on FGM in chapter 9, *The God Delusion* (Bantam Press paperback edition, 2006), pp. 369–70.

Richard Dawkins, *The Genius of Charles Darwin*, Channel 4, first screened August 2008; Richard Dawkins, 'Letter to *New Scientist* on Royal Society Row', *New Scientist*, 18 September 2008; *Horizon: A War on Science*, BBC2, first screened January 2006; for related poll results see http://www.bbc.co.uk/pressoffice/pressreleases/stories/2006/01_january/26/horizon.shtml and http://news.bbc.co.uk/1/hi/sci/tech/4648598.stm

Richard Dawkins, *The Ancestor's Tale: a Pilgrimage to the Dawn of Life* (Weidenfeld & Nicolson, 2004); the mission statement for the Richard Dawkins Foundation for Science and Reason can be found at http://richarddawkinsfoundation.org/foundation,ourMission

Stephen Jay Gould, *Rock of Ages: Science and Religion in the Fullness of Life* (Ballantine, 1999); Richard Dawkins, 'Letter to *New Scientist* on Royal Society Row', *New Scientist*, 18 September 2008. Dawkins argues that there is almost certainly no God or probably no God – for examples of this see http://www.huffingtonpost.com/richard-dawkins/why-there-almost-certainl_b_32164.html or reports on the Atheist Bus Campaign, details of which can be found at http://www.atheistbus.org.uk/; Alister McGrath, *Dawkins' God: Genes, Memes, and the Meaning of Life* (Wiley-Blackwell, 2004); for an example of the leprechaun quote see Richard Dawkins, 'Do you have to read up on leprechology before disbelieving in them?' *Independent*, 17 September 2007; For the results of the Theos thinktank poll: Caroline Lawes, *Rescuing Darwin Faith and Darwin: Harmony, Conflict, or Confusion?* (2009) can be found at http://campaigndirector.moodia.com/Client/Theos/Files/FaithandDarwin.pdf

See also Jon D. Miller, Eugenie C. Scott and Shinji Okamot, 'Public Acceptance of Evolution', *Science*, 11 August 2006, vol. 313, no. 5788, pp. 765–6; Salman Hameed, 'Bracing for Islamic Creationism', *Science*, 12 December 2008, vol. 322, no. 5908, pp. 1637–8; for a more comprehensive poll of attitudes in the US see 'You Say You Want an Evolution?: A Role for Scientists in Science Education', *Coalition of Scientific Societies*, December 2007, available online at http://opa.faseb.org/pages/PolicyIssues/sciencecoalition.htm

Owen Bennett-Jones' interview with Richard Dawkins – *Richard Dawkins on Charles Darwin*, 14 February 2009 – can be found at http://news. bbc.co.uk/1/hi/sci/tech/7885670.stm; see also Thomas Dixon, *Science and Religion: A Very Short Introduction* (Oxford University Press, 2008).

Chapter 7 and Chapter 8

See Richard Dawkins, preface of *The Blind Watchmaker* (Longman, 1986); Richard Dawkins, chapter 1.2, 'What is True?', in *A Devil's Chaplain: Reflections on Hope, Lies, Science, and Love* (Mariner Books, 2004); a full transcript of Steve Jones and Richard Dawkins speaking at the 2009 Open University annual lecture on 17 March 2009 can be found at http://www.open2.net/ dawkins/; Karl Popper, *The Logic of Scientific Discovery* (translation of *Logik der Forschung*, 1935), (Hutchinson, 1959); Karl Popper, *Unended Quest: An Intellectual Autobiography* (Fontana, 1976), pp. 171–2; Karl Popper, 'Natural selection and the emergence of mind', *Dialectica* (1978), vol. 32, pp. 339–55; Karl Popper, *Conjectures and Refutations: The Growth of Scientific Knowledge* (Basic Books, 1963), pp. 33–9, third revised edition; T.S. Kuhn, *The Structure of Scientific Revolutions* (Chicago University Press, 1962). For further reading see also Ian Hacking, *The Social Construction of What?* (Harvard University Press, 1999); Alan Chalmers, *What Is This Thing Called Science?* (Open University Press, 1999), third edition; Steven French, *Science: Key Concepts in Philosophy* (Continuum International Publishing Group Ltd, 2007).

Chapter 9

See Richard Dawkins, chapter 9, 'Selfish DNA, Jumping Genes and a Lamarckian Scare' and preface (1981) in *The Extended Phenotype: The Long Reach of the Gene* (Oxford Paperbacks, 1999), second revised edition; comments on critic in Richard Dawkins, 'I'm an atheist, BUT ...' article which can be found at http://richarddawkins.net/articleComments,318,Im-an-atheist-BUT---,Richard-Dawkins,page2; Richard Dawkins, 'The Neville

Chamberlain School of Evolutionists' in chapter 2, *The God Delusion* (Bantam Press, 2006); Richard Dawkins, *The Enemies of Reason*, Channel 4, first screened August 2007; Peter Miller, 'The Gullible Age', *Sunday Times*, 5 August 2007; Richard Dawkins, 'Extended Phenotype – but not *too* extended. A reply to Laland, Turner and Jablonka', *Biology and Philosophy* (2004), vol. 19, pp. 377–96; Richard Dawkins, preface and chapter 11, 'Doomed Rivals', in *The Blind Watchmaker* (Longman, 1986); Alfred Russel Wallace, *Darwinism: An Exposition of the Theory of Natural Selection with some of its Applications* (Macmillan and Co, 1889); George John Romanes, *Darwin and After Darwin: An Exposition of the Darwinian Theory and a Discussion of Post-Darwinian Questions* (Longmans, Green & Co, 1895), vol. II; Alfred Russel Wallace, 'Romanes *versus* Darwin', *Fortnightly Review* (1886), vol. 46, pp. 300–16; Herbert Spencer, 'The Inadequacy of Natural Selection', *The Contemporary Review* (1893), vol. 63, pp. 153–66 and 439–56; T.D.A. Cockerell, 'Recollections of Dr. Alfred Russel Wallace', *Science*, 19 December 1913, vol. 38, no. 990, pp. 871–7; *New Scientist* cover, 'Darwin Was Wrong: cutting down the tree of life', 24 January 2009; the mission statement for the Richard Dawkins Foundation for Science and Reason can be found at http://richarddawkinsfoundation.org/foundation,ourMission; 'You Say You Want an Evolution?: A Role for Scientists in Science Education', *Coalition of Scientific Societies*, December 2007, available online at http://opa.faseb.org/pages/PolicyIssues/sciencecoalition.htm; interview with Decca Aitkenhead, 'People Say I Am Strident', *Guardian*, 25 October 2008.

Conclusion

Richard Dawkins, *The Oxford Book of Modern Science Writing* (Oxford University Press, 2008); Anila Asghar, Jason R. Wiles and Brian Alters, 'Discovering International Perspectives on Biological Evolution Across Religions and Cultures; insights gained through developing methodological tools for research in diverse contexts', *International Journal of the Diversity in Organisations, Communities and Nations* (2007), vol. 6, no 4; for more details on the Dover case see Lauri Lebo, *The Devil in Dover: An Insider's Story of Dogma v. Darwin in Small-Town America* (New Press, 2008); details of the clergy letter project can be found at http://www.butler.edu/clergyproject/Backgd_info.htm; the full text of President Barack Obama's speech can be found at http://www.huffingtonpost.com/2009/05/17/obama-notre-dame-speech-f_n_204387.html